The Math Mechanic
Fractions & Mixed Numbers Edition

by

The Math Mechanic Series

First Edition

Published by Aventine Press, LLC
45 East Flower Street, Ste. 236
Chula Vista, CA 91910-7631, USA

www.aventinepress.com

ISBN: 1-59330-034-4

Printed in the United States of America

The Math Mechanic Series: Fractions & Mixed Numbers Edition
Table of Contents - Pg. 4

The Math Mechanic Series: Fractions & Mixed Numbers Edition
Table of Contents - Pg. 5

What is a Fraction?

A Fraction is part of a whole or group. A Fraction is made up of a Numerator and a Denominator. The Numerator is the number above the bar and indicates the number of parts that are selected. The Denominator is the number below the bar and indicates the total number of parts.

For example the fraction $\frac{4}{10}$ means 4 parts out of 10.

What is an Improper Fraction?

When a Fraction's Numerator is 'Larger Than' or 'Equal To' its Denominator, it is called an Improper Fraction.

Examples of this are $\frac{8}{5}$ and $\frac{2}{2}$.

What is a Mixed Number?

A Mixed Number is a combination of a Whole Number and a Fraction. An example of this is $2\frac{7}{8}$.

2 is the Whole Number, 7 is the Numerator part of the fraction, and 8 is the Denominator part of the fraction.

Diagram of a Fraction

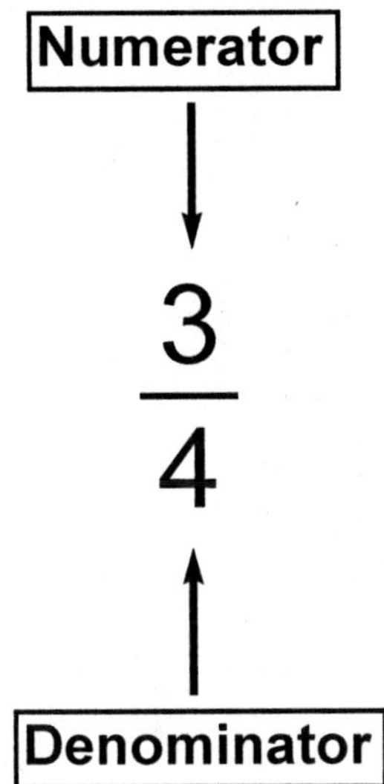

Diagram of a Mixed Number

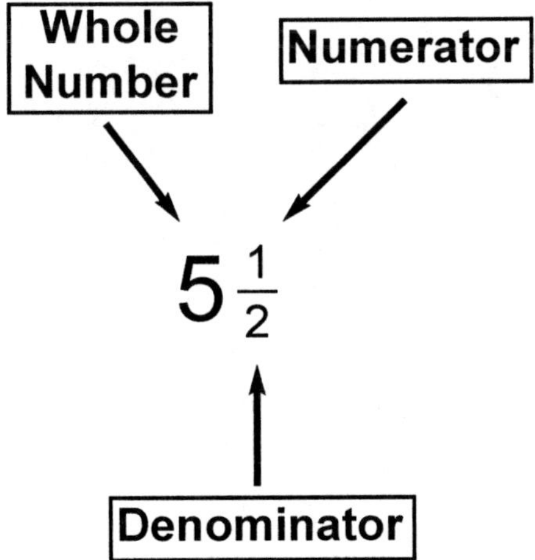

Write the Numerator of the corresponding fraction on a separate sheet of paper.

1. $\frac{1}{2}$

2. $\frac{6}{9}$

3. $\frac{3}{5}$

4. $\frac{8}{10}$

5. $\frac{5}{14}$

Write the Denominator of the corresponding fraction on a separate sheet of paper.

1. $\frac{4}{8}$

2. $\frac{2}{3}$

3. $\frac{1}{7}$

4. $\frac{15}{20}$

5. $\frac{8}{10}$

Below examples of fractions in their numerical, word, and illustrative forms.

$\frac{1}{4}$

One-fourth

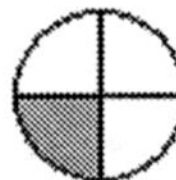

1 out of the 4 pie slices is shaded.

$\frac{2}{4}$

Two-fourths

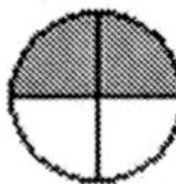

2 out of the 4 pie slices are shaded.

$\frac{3}{5}$

Three-fifths

3 out of the 5 pie slices are shaded.

$\frac{3}{4}$

Three-fourths

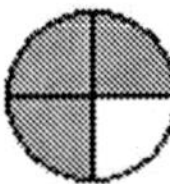

3 out of the 4 pie slices are shaded.

$\frac{2}{3}$

Two-Thirds

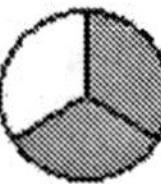

2 out of the 3 pie slices are shaded.

$\frac{4}{10}$

Four-tenths

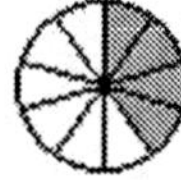

4 out of the 10 pie slices are shaded.

$\frac{8}{11}$

Eight-elevenths

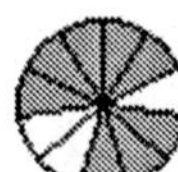

8 out of the 11 pie slices are shaded.

$\frac{4}{6}$

Four-sixths

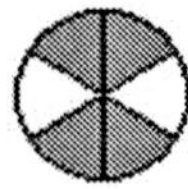

4 out of the 6 pie slices are shaded.

$\frac{7}{16}$

Seven-sixteenths

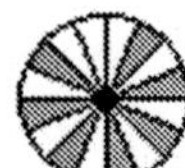

7 out of the 16 pie slices are shaded.

The Math Mechanic Series: Fractions & Mixed Numbers Edition
Fractions and Objects - Pg. 10

Below are completed examples of identifying a fraction within a collection of objects.

Write the fraction that represents the Black Squares in the set below.

There are a total of 3 squares above, 2 of which are Black.

The Black squares account for $\frac{2}{3}$ of the total number of squares.

Write the fraction that represents the Black Squares in the set below.

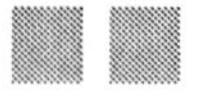 

There are a total of 5 squares above, 3 of which are Black.

The Black squares account for $\frac{3}{5}$ of the total number of squares.

Write the fraction that represents the Gray Squares in the set below.

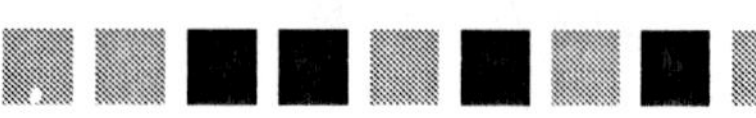

There are a total of 9 squares above, 5 of which are Gray.

The Gray squares account for $\frac{5}{9}$ of the total number of squares.

Write the fraction that represents the Gray Square in the set below.

There are a total of 4 squares above, 1 of which is Gray.

The Gray square accounts for $\frac{1}{4}$ of the total number of squares.

Write the fraction that represents the Gray Squares in the set below.

There are a total of 5 squares above, 4 of which are Gray.

The Gray squares account for $\frac{4}{5}$ of the total number of squares.

Write the fraction that represents the Black Squares in the set below.

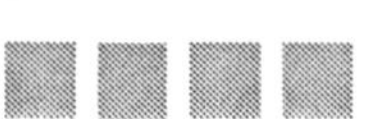 

There are a total of 7 squares above, 3 of which are Black.

The Black squares account for $\frac{3}{7}$ of the total number of squares.

The Math Mechanic Series: Fractions & Mixed Numbers Edition
Comparing Fractions Visually - Pg. 11

Below are completed examples of comparing fractional pie charts. The fractional pie charts are accompanied with their numerical and word forms.

$\frac{1}{3}$	$\frac{6}{8}$
One-third	Six-eighths
	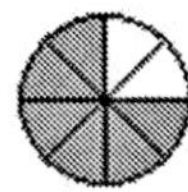

The Pie on the right is Greater Than the Pie on the left because more of its area is shaded.

$\frac{4}{7}$	$\frac{1}{9}$
Four-sevenths	One-nineth
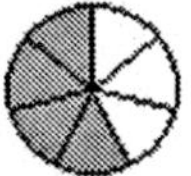	

The Pie on the left is Greater Than the Pie on the right because more of its area is shaded.

$\frac{3}{4}$	$\frac{2}{3}$
Three-fourths	Two-thirds
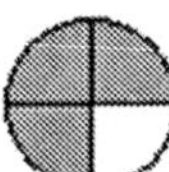	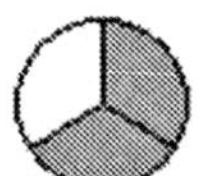

The Pie on the left is Greater Than the Pie on the right because more of its area is shaded.

$\frac{4}{8}$	$\frac{1}{2}$
Four-eighths	One-half
	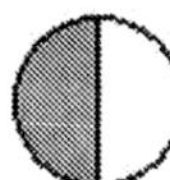

The Pie on the left is Equal To the Pie on the right because both pies have an equal amount of area shaded.

$\frac{5}{6}$	$\frac{7}{8}$
Five-sixths	Seven-eighths
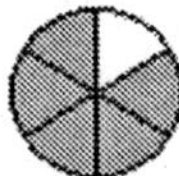	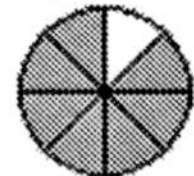

The Pie on the right is Greater Than the Pie on the left because more of its area is shaded.

$\frac{1}{3}$	$\frac{3}{9}$
One-third	Three-nineths

The Pie on the left is Equal To the Pie on the right because both pies have an equal amount of area shaded.

You can compare fractions by taking each fraction and dividing the Numerator by the Denominator. The fraction whose division yields the larger result is the larger fraction.

Example: Compare $\frac{1}{2}$ and $\frac{2}{3}$.

$1 \div 2 = 0.50$

$2 \div 3 = 0.67$

0.67 is larger than 0.50, therefore $\frac{2}{3}$ **is larger than** $\frac{1}{2}$.

Example: Compare $\frac{4}{5}$ and $\frac{3}{4}$.

$4 \div 5 = 0.80$

$3 \div 4 = 0.75$

0.80 is larger than 0.75, therefore $\frac{4}{5}$ **is larger than** $\frac{3}{4}$.

Example: Compare $\frac{5}{7}$ and $\frac{3}{4}$.

$5 \div 7 = 0.71$

$3 \div 4 = 0.75$

0.75 is larger than 0.71, therefore $\frac{3}{4}$ **is larger than** $\frac{5}{7}$.

Example: Compare $\frac{6}{8}$ and $\frac{6}{10}$.

$6 \div 8 = 0.75$

$6 \div 10 = 0.60$

0.75 is larger than 0.60, therefore $\frac{6}{8}$ **is larger than** $\frac{6}{10}$.

The procedures for converting a Fraction to a Percent are below.

Step 1. Divide the fraction's Numerator by the Denominator.
Step 2. Multiply the result of the division by 100.
Step 3A. If there is to be no decimal place, you will round to the nearest one.
Step 3B. If the Percent is to include a decimal place, you will round the decimal up or down to the instructed decimal place.
Step 4. Attach the percent sign ('%') to the end.

Example: Convert $\frac{4}{5}$ to a percent with no decimal place.

1. $4 \div 5 = 0.80$
2. 0.80 x 100 = 80.0
3A. There will be no decimal place so 80.0 will be rounded to the nearest one. 80.0 rounded to the nearest one is 80
4. 80 becomes 80% after the '%' sign is attached.

$\mathbf{\frac{4}{5}}$ **converted to a percent is 80%**

Example: Convert $\frac{3}{9}$ to a percent with one decimal place.

1. $3 \div 9 = 0.333333333$
2. 0.333333333 x 100 = 33.3333333
3B. There will be one decimal place so 33.3333333 will be rounded to the nearest tenth.
33.3333333 rounded to the nearest tenth is 33.3
4. 33.3 becomes 33.3% after the '%' sign is attached.

$\mathbf{\frac{3}{9}}$ **converted to a percent is 33.3%**

Example: Convert $\frac{5}{8}$ to a percent with two decimal places.

1. $5 \div 8 = 0.625$
2. 0.625 x 100 = 62.5
3B. There will be two decimal places so 62.5 will be rounded to the nearest hundredth. 62.5 rounded to the nearest hundredth is 62.50
4. 62.50 becomes 62.50% after the '%' sign is attached.

$\mathbf{\frac{5}{8}}$ **converted to a percent is 62.50%**

The procedures for converting a Percent to a Fraction are below.

Step 1. Remove the '%' sign from the percent.
Step 2. Change the Percent to a Fraction by placing the Percent over 100.
The Percent becomes the Numerator and 100 becomes the Denominator.

Example: Convert 25% to a fraction.

1. 25% becomes 25 after the '%' sign is removed.
2. Place 25 over 100

$$\frac{25}{100}$$

25% converted to a Fraction is $\frac{25}{100}$

Example: Convert 48% to a fraction.

1. 48% becomes 48 after the '%' sign is removed.
2. Place 48 over 100

$$\frac{48}{100}$$

48% converted to a Fraction is $\frac{48}{100}$

Example: Convert 61% to a fraction.

1. 61% becomes 61 after the '%' sign is removed.
2. Place 61 over 100

$$\frac{61}{100}$$

61% converted to a Fraction is $\frac{61}{100}$

The procedures for converting a fraction to a decimal are below.

Step 1. Divide the fraction's Numerator by the Denominator.
Step 2. Round the decimal up or down to the instructed decimal place.

Example: Convert the Fraction $\frac{6}{7}$ to a Decimal. Round the Decimal to the nearest tenth.

1. 6 ÷ 7 = 0.8571428571429
2. 0.8571428571429 rounded to the nearest tenth is 0.9

$\frac{6}{7}$ **converted to a Decimal rounded to the nearest tenth is 0.9**

Example: Convert the Fraction $\frac{1}{4}$ to a Decimal. Round the Decimal to the nearest hundredth.

1. 1 ÷ 4 = 0.25
2. 0.25 rounded to the nearest hundredth is 0.25

$\frac{1}{4}$ **converted to a Decimal rounded to the nearest hundredth is 0.25**

Example: Convert the Fraction $\frac{4}{7}$ to a Decimal. Round the Decimal to the nearest hundredth.

1. 4 ÷ 7 = 0.571428571428
2. 0.571428571428 rounded to the nearest hundredth is 0.57

$\frac{4}{7}$ **converted to a Decimal rounded to the nearest hundredth is 0.57**

Example: Convert the Fraction $\frac{10}{35}$ to a Decimal. Round the Decimal to the nearest tenth.

1. 10 ÷ 35 = 0.285714
2. 0.285714 rounded to the nearest tenth is 0.3

$\frac{10}{35}$ **converted to a Decimal rounded to the nearest tenth is 0.3**

The procedure for converting a Non-Repeating Decimal to a fraction is below.
Procedure: Take the value of the decimal and place it over the appropriate denominator.

For Example 0.4 is the same as "Four Tenths", therefore to convert .4 to a fraction you would place 4 over 10. $\frac{4}{10}$

More Examples Below

Example: Convert 0.6 to a Fraction.
0.6 is the same as "Six Tenths", therefore place 6 over 10 to convert it to a Fraction.

0.6 converted to a Fraction is $\frac{6}{10}$

Example: Convert 0.35 to a Fraction.
0.35 is the same as "Thirty-five Hundredths", therefore place 35 over 100 to convert it to a Fraction.

0.35 converted to a Fraction is $\frac{35}{100}$

Example: Convert 0.114 to a Fraction.
0.114 is the same as "One Hundred Fourteen Thousandths", therefore place 114 over 1000 to convert it to a Fraction.

0.114 converted to a Fraction is $\frac{114}{1000}$

Example: Convert 0.08 to a Fraction.
0.08 is the same as "Eight Hundredths", therefore place 8 over 100 to convert it to a Fraction.

0.08 converted to a Fraction is $\frac{8}{100}$

When there is a number in front of the decimal point, then the decimal will become a Mixed Number when converted. The number in front of the decimal point becomes the Whole Number part and the number after the decimal point becomes the Numerator.

Example: Convert 1.9 to a Mixed Number.
1.9 is the same as "One and Nine Tenths", therefore 1 becomes the Whole Number part of the Mixed Number. Place 9 over 10 to create the Fraction part of the Mixed Number.

1.9 converted to a Mixed Number is $1\frac{9}{10}$

Example: Convert 5.37 to a Mixed Number.
5.37 is the same as "Five and Thirty-seven Hundredths", therefore 5 becomes the Whole Number part of the Mixed Number. Place 37 over 100 to create the Fraction part of the Mixed Number.

5.37 converted to a Mixed Number is $5\frac{37}{100}$

The procedures for converting a Repeating Decimal to a Fraction are below.

Step 1. Create an equation in the form "Variable" = "Decimal Number". This will be the original equation.
Example: x = 0.33333

Step 2. Multiply the statement by a power of ten. You will use the power of ten that reflects the amount of digits that repeat in the Decimal Number.

For Instance

If it contains 1 repeating digit, you will multiply by 10 since there is one zero in ten.
If it contains 2 repeating digits, you will multiply by 100, since there are 2 zeroes in 100.
If it contains 3 repeating digits, you will multiply by 1000, since there are 3 zeroes in 1000.
Follow the same procedure for Decimals that repeat every 4, 5, 6, digits and so forth.

Example: 0.33333 contains one repeating digit, so you would multiply x = 0.33333 by 10.
$10 \cdot x = 10 \cdot 0.33333$ equals 10x = 3.3333

Example: 0.757575 contains two repeating digits, so you would multiply x = 0.757575 by 100.
$100 \cdot x = 100 \cdot 0.757575$ equals 100x = 75.7575

Example: 0.451451 contains three repeating digits, so you would multiply x = 0.451451 by 1000.
$1000 \cdot x = 1000 \cdot 0.451451$ equals 1000x = 451.451

Step 3. Subtract the original equation from the multiplied equation in Step 2.

Example:
$$\begin{array}{r} 10x = 3.3333 \\ \underline{-\quad x = 0.33333} \\ 9x = 2.9997 \end{array}$$

Step 4. Round the decimal number on the right side of the equation **(right of the '=')** to the nearest One.
Example: 9x = 2.9997 changes to 9x = 3 after the right side has been rounded to the nearest One.

Step 5. Isolate the x on the left side by removing the coefficient **(number attached to the x on the left side)**.
This is done by dividing both sides by the coefficient.

Example: $\frac{9x}{9} = \frac{3}{9}$ equals $x = \frac{3}{9}$

The x has been isolated and the right side has become a Fraction. This Fraction is the answer to the conversion problem.

Example: Convert 0.777777 to a Fraction.

1. $x = 0.777777$
2. $10 \cdot x = 10 \cdot 0.777777$ equals $10x = 7.77777$

3.
$$\begin{array}{r} 10x = 7.77777 \\ -\ \ x = 0.777777 \\ \hline 9x = 6.999993 \end{array}$$

4. $9x = 7$

5. $\frac{9x}{9} = \frac{7}{9}$ equals $x = \frac{7}{9}$

0.777777 converted to a Fraction is $\frac{7}{9}$

Example: Convert 0.232323 to a Fraction.

1. $x = 0.232323$
2. $100 \cdot x = 100 \cdot 0.232323$ equals $100x = 23.2323$

3.
$$\begin{array}{r} 100x = 23.2323 \\ -\ \ x = 0.232323 \\ \hline 99x = 22.999977 \end{array}$$

4. $99x = 23$

5. $\frac{99x}{99} = \frac{23}{99}$ equals $x = \frac{23}{99}$

0.232323 converted to a Fraction is $\frac{23}{99}$

Example: Convert 1.456456 to a Mixed Number.

1. $x = 1.456456$
2. $1000 \cdot x = 1000 \cdot 1.456456$ equals $1000x = 1456.456$

3.
$$\begin{array}{r} 1000x = 1456.456 \\ -\ \ x = 1.456456 \\ \hline 999x = 1454.999544 \end{array}$$

4. $999x = 1455$

5. $\frac{999x}{999} = \frac{1455}{999}$ equals $x = \frac{1455}{999}$

$\frac{1455}{999}$ converted to a Mixed Number is $1\frac{456}{999}$.

1.456456 converted to a Mixed Number is $1\frac{456}{999}$

There will be times when you will need to convert a Whole Number to an Improper Fraction.

The procedure to convert a Whole Number to an Improper Fraction is as follows:

Take the Whole Number and place it over a bar, then place the number 1 under the bar.

Note: The Whole Number becomes the Numerator of the improper fraction and the number 1 becomes the Denominator.

Example: Convert 3 to an improper fraction.

Place 3 over a bar, then place the number 1 under the bar.

$$3 = \frac{3}{1}$$

Example: Convert 49 to an improper fraction.

Place 49 over a bar, then place the number 1 under the bar.

$$49 = \frac{49}{1}$$

Example: Convert 128 to an improper fraction.

Place 128 over a bar, then place the number 1 under the bar.

$$128 = \frac{128}{1}$$

Example: Convert 4,479 to an improper fraction.

Place 4,479 over a bar, then place the number 1 under the bar.

$$4{,}479 = \frac{4{,}479}{1}$$

Convert 4 to an improper fraction. $4 = \frac{4}{1}$	To convert 4 to an improper fraction, place 4 over a bar, then place the number 1 under the bar. 4 becomes the Numerator of the improper fraction and 1 becomes the Denominator.

Convert 10 to an improper fraction. $10 = \frac{10}{1}$	To convert 10 to an improper fraction, place 10 over a bar, then place the number 1 under the bar. 10 becomes the Numerator of the improper fraction and 1 becomes the Denominator.

Convert 7 to an improper fraction. $7 = \frac{7}{1}$	To convert 7 to an improper fraction, place 7 over a bar, then place the number 1 under the bar. 7 becomes the Numerator of the improper fraction and 1 becomes the Denominator.

Convert 5 to an improper fraction. $5 = \frac{5}{1}$	To convert 5 to an improper fraction, place 5 over a bar, then place the number 1 under the bar. 5 becomes the Numerator of the improper fraction and 1 becomes the Denominator.

Converting Improper Fractions To Whole Numbers - Pg. 21

You can convert an Improper Fraction to a Whole Number if:
The Denominator will divide into the Numerator without a remainder.

If there is a remainder, then the improper fraction cannot be converted to a Whole Number. It can however be converted to a Mixed Number. This is because the remainder becomes the Numerator part of the Mixed Number.

$\frac{8}{4}$ can be converted to a Whole Number since there is no remainder when dividing 8 by 4.

$8 \div 4 = 2$

$\frac{9}{4}$ cannot be converted to a Whole Number since there is a remainder when dividing 9 by 4.

$9 \div 4 = 2$ with a Remainder of 1

More Examples Below

Example: Convert $\frac{10}{5}$ to a Whole Number.

Divide the Numerator by the Denominator. $10 \div 5 = 2$
2 becomes the Whole Number.

Example: Convert $\frac{6}{1}$ to a Whole Number.

Divide the Numerator by the Denominator. $6 \div 1 = 6$
6 becomes the Whole Number.

Example: Convert $\frac{40}{4}$ to a Whole Number.

Divide the Numerator by the Denominator. $40 \div 4 = 10$
10 becomes the Whole Number.

Example: Convert $\frac{14}{2}$ to a Whole Number.

Divide the Numerator by the Denominator. $14 \div 2 = 7$
7 becomes the Whole Number.

Example: Convert $\frac{30}{6}$ to a Whole Number.

Divide the Numerator by the Denominator. $30 \div 6 = 5$
5 becomes the Whole Number.

Example: Convert $\frac{8}{8}$ to a Whole Number.

Divide the Numerator by the Denominator. $8 \div 8 = 1$
1 becomes the Whole Number.

Convert $\frac{24}{3}$ to a Whole Number.

1. 24 ÷ 3 = 8

$\frac{24}{3}$ **converts to 8.**

Divide the Numerator by the Denominator.

24 ÷ 3 = 8

Convert $\frac{7}{1}$ to a Whole Number.

1. 7 ÷ 1 = 7

$\frac{7}{1}$ **converts to 7.**

Divide the Numerator by the Denominator.

7 ÷ 1 = 7

Convert $\frac{15}{5}$ to a Whole Number.

1. 15 ÷ 5 = 3

$\frac{15}{5}$ **converts to 3.**

Divide the Numerator by the Denominator.

15 ÷ 5 = 3

Convert $\frac{21}{1}$ to a Whole Number.

1. 21 ÷ 1 = 21

$\frac{21}{1}$ **converts to 21.**

Divide the Numerator by the Denominator.

21 ÷ 1 = 21

Convert $\frac{52}{13}$ to a Whole Number.

1. 52 ÷ 13 = 4

$\frac{52}{13}$ **converts to 4.**

Divide the Numerator by the Denominator.

52 ÷ 13 = 4

There will be times you when you will need to convert a Mixed Number to an Improper Fraction.

The procedures to convert a Mixed Number to an Improper Fraction are below:

Step 1. Multiply the Whole Number by the Denominator.
Note: Once this happens, the Whole Number will no longer exist.
Step 2. Add the Numerator to the result of the previous multiplication **(Step 1)**.
Note: The result becomes the new Numerator.

The Denominator stays the same.

Example: Convert $2\frac{7}{8}$ to an Improper Fraction.

Step 1. 2 x 8 = 16
Step 2. 16 + 7 = 23

$\mathbf{2\frac{7}{8}}$ **converts to** $\mathbf{\frac{23}{8}}$

Example: Convert $39\frac{1}{3}$ to an Improper Fraction.

Step 1. 39 x 3 = 117
Step 2. 117 + 1 = 118

$\mathbf{39\frac{1}{3}}$ **converts to** $\mathbf{\frac{118}{3}}$ **.**

Example: Convert $4\frac{2}{5}$ to an Improper Fraction.

Step 1. 4 x 5 = 20
Step 2. 20 + 2 = 22

$\mathbf{4\frac{2}{5}}$ **converts to** $\mathbf{\frac{22}{5}}$

Convert $1\frac{4}{5}$ to an improper fraction.

1. 1 x 5 = 5

2. 5 + 4 = 9

$1\frac{4}{5}$ converts to $\frac{9}{5}$

1. Multiply the Whole Number by the Denominator.
The Whole Number part of the Mixed Number is 1 and the fraction's Denominator is 5.

1 x 5 = 5

2. Add the Numerator to the result of the previous multiplication. The Numerator is 4 and the result of the previous multiplication is 5 **(Step 1)**.

5 + 4 = 9

9 becomes the new Numerator.

The Denominator which is 5 will stay the same.

Convert $4\frac{3}{7}$ to an improper fraction.

1. 4 x 7 = 28

2. 28 + 3 = 31

$4\frac{3}{7}$ converts to $\frac{31}{7}$

1. Multiply the Whole Number by the Denominator.
The Whole Number part of the Mixed Number is 4 and the fraction's Denominator is 7.

4 x 7 = 28

2. Add the Numerator to the result of the previous multiplication. The Numerator is 3 and the result of the previous multiplication is 28 **(Step 1)**.

28 + 3 = 31

31 becomes the new Numerator.

The Denominator which is 7 will stay the same.

There will be times you when you will need to convert an Improper Fraction to a Mixed Number.

The procedure to convert an Improper Fraction to a Mixed Number is as follows:
Divide the Numerator by the Denominator. The quotient becomes the Whole Number part of the Mixed Number and the remainder becomes the Numerator.

Note: The Denominator stays the same.

Example: Convert $\frac{9}{5}$ to a Mixed Number.

$9 \div 5 = 1$ with a Remainder of 4
1 becomes the Whole Number part of the Mixed Number and 4 becomes the Numerator. The Denominator stays the same. The Denominator in this case is 5.

$1\frac{4}{5}$

$\frac{9}{5}$ converts to $1\frac{4}{5}$

Example: Convert $\frac{19}{6}$ to a Mixed Number.

$19 \div 6 = 3$ with a Remainder of 1
3 becomes the Whole Number part of the Mixed Number and 1 becomes the Numerator. The Denominator stays the same. The Denominator in this case is 6.

$3\frac{1}{6}$

$\frac{19}{6}$ converts to $3\frac{1}{6}$

Convert $\frac{7}{2}$ to a Mixed Number.

$7 \div 2 = 3$ with a Remainder of 1

$3\frac{1}{2}$

$\frac{7}{2}$ converts to $3\frac{1}{2}$

Divide the Numerator by the Denominator.
$7 \div 2 = 3$ with a Remainder of 1

The Quotient becomes the Whole Number part of the Mixed Number and the remainder becomes the Numerator. The Quotient is 3 and the Remainder is 1.

3 becomes the Whole Number part of the Mixed Number.

1 becomes the Numerator of the fraction part.

The Denominator stays the same. In this case the Denominator is 2.

Convert $\frac{15}{6}$ to a Mixed Number.

$15 \div 6 = 2$ with a Remainder of 3

$2\frac{3}{6}$

$\frac{15}{6}$ converts to $2\frac{3}{6}$

Divide the Numerator by the Denominator.
$15 \div 6 = 2$ with a Remainder of 3

The Quotient becomes the Whole Number part of the Mixed Number and the remainder becomes the Numerator. The Quotient is 2 and the Remainder is 3.

2 becomes the Whole Number part of the Mixed Number.

3 becomes the Numerator of the fraction part.

The Denominator stays the same. In this case the Denominator is 6.

Convert $\frac{55}{12}$ to a Mixed Number.

$55 \div 12 = 4$ with a Remainder of 7

$4\frac{7}{12}$

$\frac{55}{12}$ converts to $4\frac{7}{12}$

Divide the Numerator by the Denominator.
$55 \div 12 = 4$ with a Remainder of 7

The Quotient becomes the Whole Number part of the Mixed Number and the remainder becomes the Numerator. The Quotient is 4 and the Remainder is 7.

4 becomes the Whole Number part of the Mixed Number.

7 becomes the Numerator of the fraction part.

The Denominator stays the same. In this case the Denominator is 12.

The Math Mechanic Series: Fractions & Mixed Numbers Edition
Retrieving the Factors of a Number - Pg. 27

The procedures for retrieving the factors of a number are below:

Step 1. Retrieve the smallest and largest factors. The smallest factor will always be 1 and the largest factor is the number that you are retrieving the factors for.

Example: 36 - The smallest factor is 1. The largest factor is 36.
109 - The smallest factor is 1. The largest factor is 109.
4 - The smallest factor is 1. The largest factor is 4.
2 - The smallest factor is 1. The largest factor is 2.
24 - The smallest factor is 1. The largest factor is 24.

Step 2. Check to see if the number that you are trying to find the factors for is an even number or an odd number.

A. If the number that you are finding the factors for is an odd number

1. Divide the number that you are finding the factors for by 2 to get the halfway point. The Quotient is the halfway point (ignore the Remainder).
Example: 39 ÷ 2 = 19 Remainder 1
19 will be the halfway point.

B. If the number that you are finding the factors for is an even number

1. Divide the number that you are finding the factors for by 2 to get the halfway point. The Quotient is the halfway point.
Example: 24 ÷ 2 = 12

2. Since there is no Remainder, the Quotient and 2 are factors of the number that you are finding factors for.

Step 3. To get the rest of the factors, you will have to divide into the number that you are trying to find the factors for starting with the number 3. If the result of the division does not have a remainder, then the divisor and quotient are factors of the number.
Always start off by dividing by 3, then increment your way up until you get to the number that equals the (halfway point - 1).

Note: When you reach the point where you attempt to divide by a number that has already been proven to be a factor, you have found all of the factors and the problem is complete.

An example of Step 3 is below using the number 24.
The halfway point of 24 is 12. Subtract 1 from the halfway point.
12 - 1 = 11, so you would divide 24 by the numbers 3 thru 11.
Example: *24 ÷ 3, 24 ÷ 4, 24 ÷ 5, 24 ÷ 6, etc.*

An example of the above Note is below using the number 24.
The halfway point of 24 is 12. Subtract 1 from the halfway point.
12 - 1 = 11, so you would divide 24 by the numbers 3 thru 11.
Example: *24 ÷ 3, 24 ÷ 4, 24 ÷ 5, 24 ÷ 6, etc.*

When 24 is divided by 4 you will get 6 as the quotient.
You would then divide 24 by 5. 24 ÷ 5 = 4 Remainder 4. 5 is not a factor of 24.
Now you will divide 24 by 6. Let us pause for a moment. We have already found that 6 is a factor of 24. This is because 24 ÷ 4 = 6.
At this point you have reached a number that has already proven to be a factor, you can now stop performing Step 3 because you have found all the factors for the given number.

To reiterate, divide the number by 3 thru the (halfway point - 1) or until you attempt to divide by a number that has been proven to be a factor. When either one of these circumstances occurs you have found all factors for the number and can stop.

Example: One less than half of 24 is 11, so start dividing 24 by the numbers 3 thru 11.

24 ÷ 3 = 8	There is No Remainder. 3 and 8 are factors of 24.
24 ÷ 4 = 6	There is No Remainder. 4 and 6 are factors of 24.
24 ÷ 5 = 4 R4	There is a Remainder of 4. 5 is not a factor of 24.
24 ÷ 6	(There is no need to proceed with this division because 6 has already proven to be a factor. We can now stop performing Step 3.)

Example: Find the Factors of 16

Step 1. Retrieve the smallest and largest factors of 16. The smallest factor is 1 and the largest factor is the number that you are retrieving the factors for.

1 is the smallest Factor
16 is the largest Factor

Step 2. Check to see if the number that you are finding the factors for is an even number or an odd number.

2B. 16 is an even number.

2B1. 16 ÷ 2 = 8 (8 is the halfway point of 16)

2B2. 2 and 8 are factors of 16

Step 3. To get the rest of the factors, you will have to divide into the number that you are trying to find the factors for starting with the number 3. If the result of the division does not have a remainder, then the divisor and quotient are factors of the number.

Divide the number by 3 thru the (halfway point - 1) or until you attempt to divide by a number that has been proven to be a factor.

The halfway point of 16 is 8. 8 - 1 = 7, so start dividing 16 by the numbers 3 thru 7.
16 ÷ 3 = 5 R1 There is a Remainder of 1. 3 is not a factor of 16.
16 ÷ 4 = 4 There is No Remainder. 4 is a factor of 16.
16 ÷ 5 = 3 R1 There is a Remainder of 1. 5 is not a factor of 16.
16 ÷ 6 = 2 R4 There is a Remainder of 4. 6 is not a factor of 16.
16 ÷ 7 = 2 R2 There is a Remainder of 2. 7 is not a factor of 16.

The Factors for the number 16 are 1, 2, 4, 8, and 16.

The Math Mechanic Series: Fractions & Mixed Numbers Edition
Step-by-Step Examples of Finding the Factors of a Number - Pg. 29

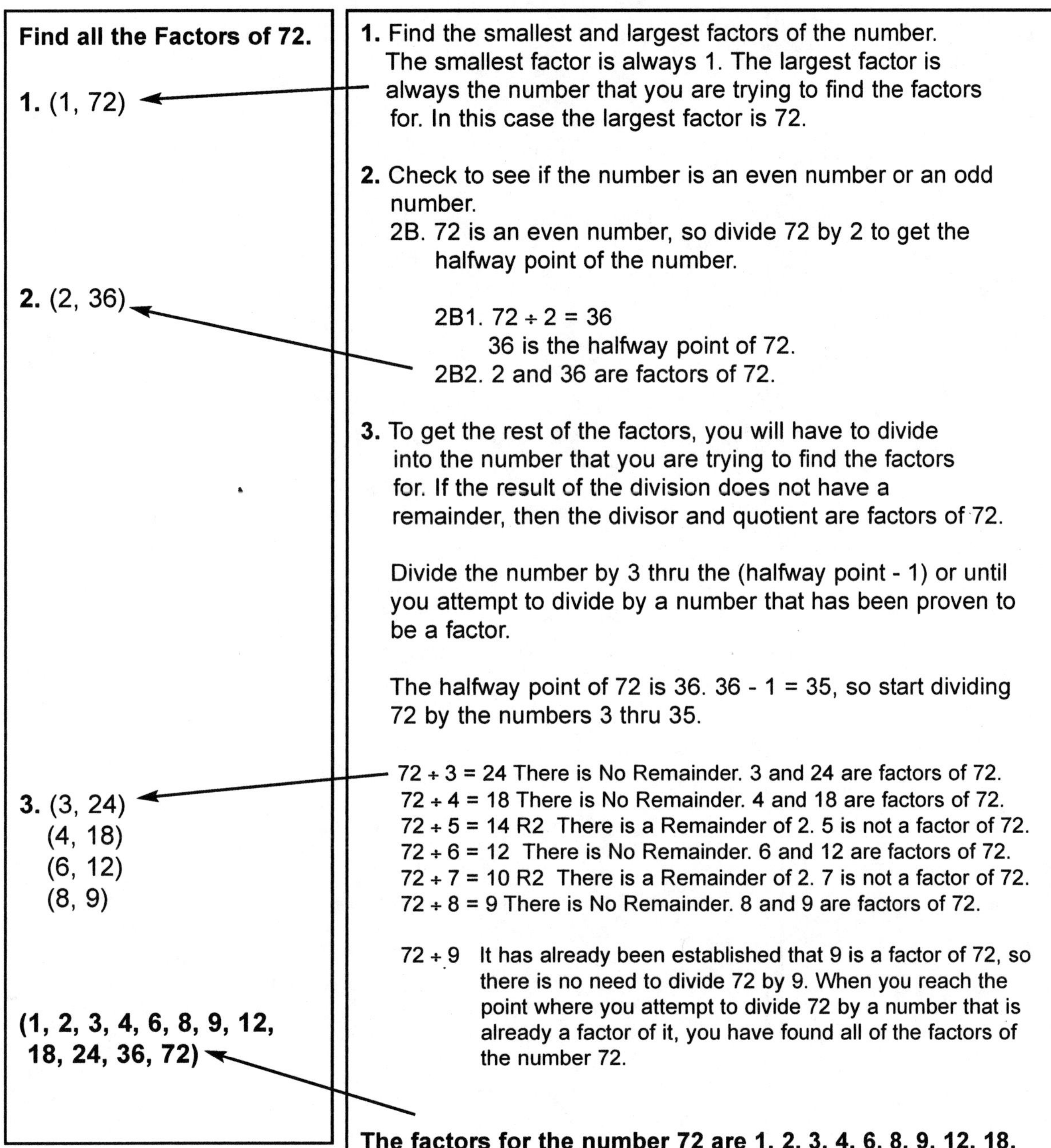

Find all the Factors of 72.

1. (1, 72)

2. (2, 36)

3. (3, 24)
(4, 18)
(6, 12)
(8, 9)

(1, 2, 3, 4, 6, 8, 9, 12, 18, 24, 36, 72)

1. Find the smallest and largest factors of the number. The smallest factor is always 1. The largest factor is always the number that you are trying to find the factors for. In this case the largest factor is 72.

2. Check to see if the number is an even number or an odd number.
2B. 72 is an even number, so divide 72 by 2 to get the halfway point of the number.

2B1. 72 ÷ 2 = 36
36 is the halfway point of 72.
2B2. 2 and 36 are factors of 72.

3. To get the rest of the factors, you will have to divide into the number that you are trying to find the factors for. If the result of the division does not have a remainder, then the divisor and quotient are factors of 72.

Divide the number by 3 thru the (halfway point - 1) or until you attempt to divide by a number that has been proven to be a factor.

The halfway point of 72 is 36. 36 - 1 = 35, so start dividing 72 by the numbers 3 thru 35.

72 ÷ 3 = 24 There is No Remainder. 3 and 24 are factors of 72.
72 ÷ 4 = 18 There is No Remainder. 4 and 18 are factors of 72.
72 ÷ 5 = 14 R2 There is a Remainder of 2. 5 is not a factor of 72.
72 ÷ 6 = 12 There is No Remainder. 6 and 12 are factors of 72.
72 ÷ 7 = 10 R2 There is a Remainder of 2. 7 is not a factor of 72.
72 ÷ 8 = 9 There is No Remainder. 8 and 9 are factors of 72.

72 ÷ 9 It has already been established that 9 is a factor of 72, so there is no need to divide 72 by 9. When you reach the point where you attempt to divide 72 by a number that is already a factor of it, you have found all of the factors of the number 72.

The factors for the number 72 are 1, 2, 3, 4, 6, 8, 9, 12, 18, 24, 36, and 72.

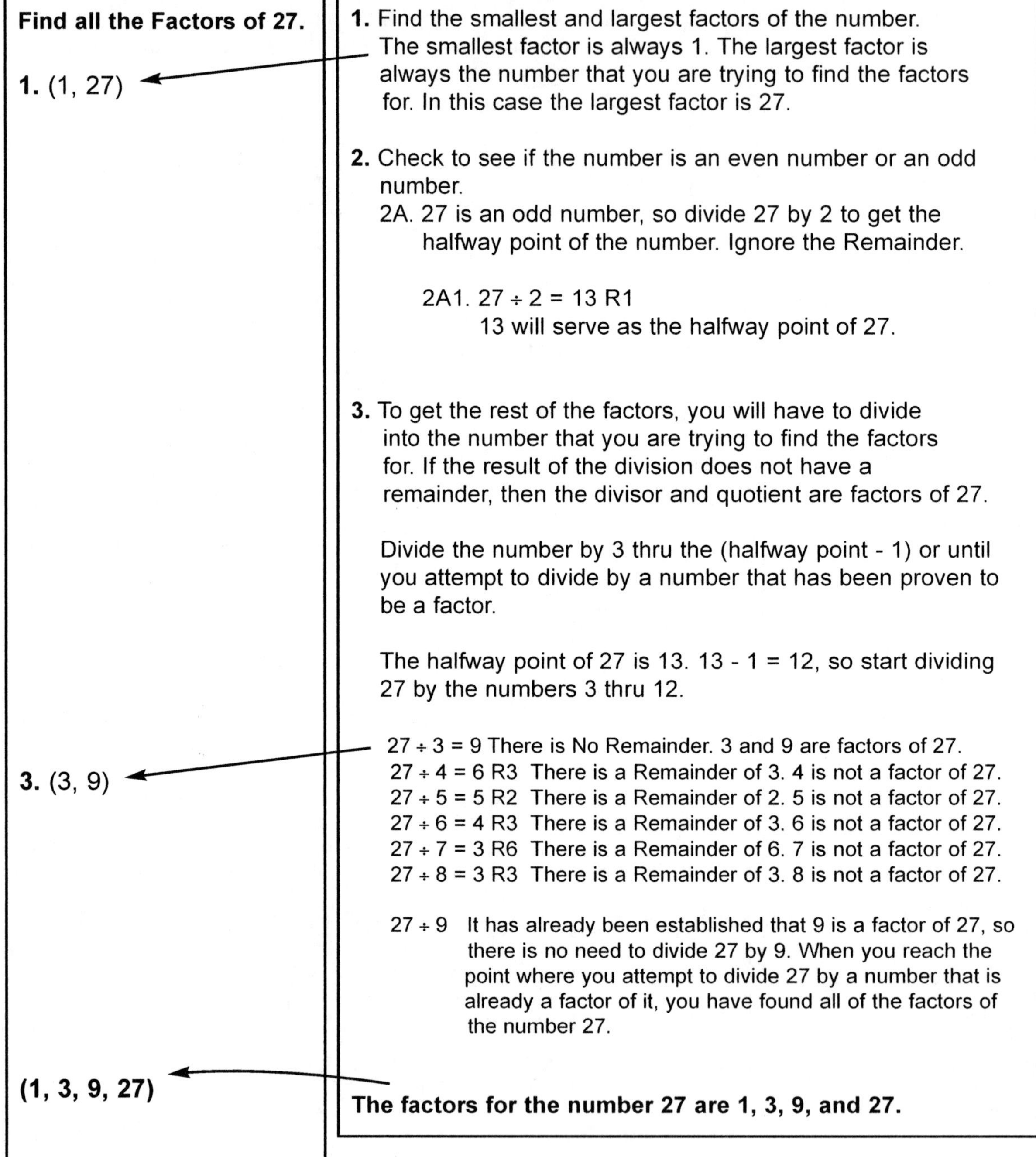

Find all the Factors of 27.

1. (1, 27)

3. (3, 9)

(1, 3, 9, 27)

1. Find the smallest and largest factors of the number. The smallest factor is always 1. The largest factor is always the number that you are trying to find the factors for. In this case the largest factor is 27.

2. Check to see if the number is an even number or an odd number.

2A. 27 is an odd number, so divide 27 by 2 to get the halfway point of the number. Ignore the Remainder.

2A1. 27 ÷ 2 = 13 R1
13 will serve as the halfway point of 27.

3. To get the rest of the factors, you will have to divide into the number that you are trying to find the factors for. If the result of the division does not have a remainder, then the divisor and quotient are factors of 27.

Divide the number by 3 thru the (halfway point - 1) or until you attempt to divide by a number that has been proven to be a factor.

The halfway point of 27 is 13. 13 - 1 = 12, so start dividing 27 by the numbers 3 thru 12.

27 ÷ 3 = 9 There is No Remainder. 3 and 9 are factors of 27.
27 ÷ 4 = 6 R3 There is a Remainder of 3. 4 is not a factor of 27.
27 ÷ 5 = 5 R2 There is a Remainder of 2. 5 is not a factor of 27.
27 ÷ 6 = 4 R3 There is a Remainder of 3. 6 is not a factor of 27.
27 ÷ 7 = 3 R6 There is a Remainder of 6. 7 is not a factor of 27.
27 ÷ 8 = 3 R3 There is a Remainder of 3. 8 is not a factor of 27.

27 ÷ 9 It has already been established that 9 is a factor of 27, so there is no need to divide 27 by 9. When you reach the point where you attempt to divide 27 by a number that is already a factor of it, you have found all of the factors of the number 27.

The factors for the number 27 are 1, 3, 9, and 27.

Step-by-Step Examples of Finding the Factors of a Number - Pg. 31

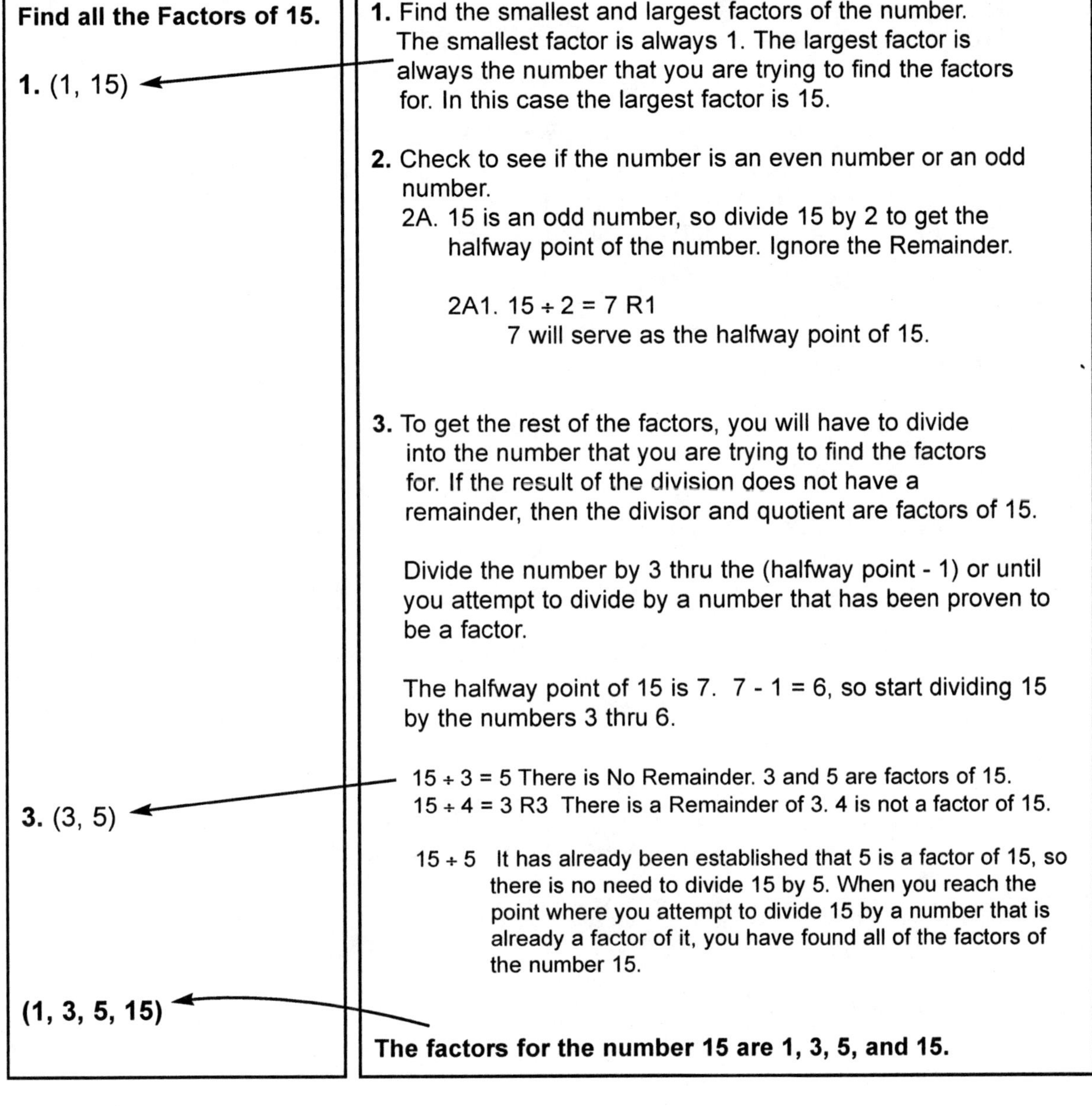

Find all the Factors of 15.

1. (1, 15)

3. (3, 5)

(1, 3, 5, 15)

1. Find the smallest and largest factors of the number. The smallest factor is always 1. The largest factor is always the number that you are trying to find the factors for. In this case the largest factor is 15.

2. Check to see if the number is an even number or an odd number.

2A. 15 is an odd number, so divide 15 by 2 to get the halfway point of the number. Ignore the Remainder.

2A1. 15 ÷ 2 = 7 R1
7 will serve as the halfway point of 15.

3. To get the rest of the factors, you will have to divide into the number that you are trying to find the factors for. If the result of the division does not have a remainder, then the divisor and quotient are factors of 15.

Divide the number by 3 thru the (halfway point - 1) or until you attempt to divide by a number that has been proven to be a factor.

The halfway point of 15 is 7. 7 - 1 = 6, so start dividing 15 by the numbers 3 thru 6.

15 ÷ 3 = 5 There is No Remainder. 3 and 5 are factors of 15.
15 ÷ 4 = 3 R3 There is a Remainder of 3. 4 is not a factor of 15.

15 ÷ 5 It has already been established that 5 is a factor of 15, so there is no need to divide 15 by 5. When you reach the point where you attempt to divide 15 by a number that is already a factor of it, you have found all of the factors of the number 15.

The factors for the number 15 are 1, 3, 5, and 15.

The procedures for finding the Greatest Common Factor of 2 or more numbers are below:

Step 1. Take the first number and find all factors of that number.
Step 2. Repeat Step 1 for the other numbers.
Step 3. Pick out all of the factors that are common in each number.
Step 4. Select the largest factor of all the common factors.

Example: Find the GCF of 18 and 30.

1. The factors of 18 are 1, 2, 3, 6, 9, and 18.
2. The factors of 30 are 1, 2, 3, 5, 6, 10, 15, and 30.
3. The factors that 18 and 30 have in common are 1, 2, 3, and 6.
4. The largest of the common factors is 6.

The GCF of 18 and 30 is 6.

Example: Find the GCF of 11 and 21.

1. The factors of 11 are 1 and 11.
2. The factors of 21 are 1, 3, 7, and 21.
3. The factor that 11 and 21 have in common is 1.
4. The only common factor is 1, therefore it is the Greatest Common Factor.

The GCF of 11 and 21 is 1.

Example: Find the GCF of 50 and 100.

1. The factors of 50 are 1, 2, 5, 10, 25, and 50.
2. The factors of 100 are 1, 2, 4, 5, 10, 20, 25, 50, and 100.
3. The factors that 50 and 100 have in common are 1, 2, 5, 10, 25, and 50.
4. The largest of the common factors is 50.

The GCF of 50 and 100 is 50.

The procedures to reduce a fraction to its Lowest Terms are below.

Step 1. Find the GCF of the Numerator and Denominator.
Note: If the GCF of the Numerator and Denominator is 1, then the fraction is already in its Lowest Terms, therefore you will not perform Step 2.

Step 2. Divide both the Numerator and Denominator by the GCF. The results of both of the divisions will replace the Numerator and Denominator respectively.

Example: Reduce $\frac{24}{32}$ to its Lowest Terms.

1. The GCF of 24 and 32 is 8.
2. $24 \div 8 = 3$ 3 replaces 24 in the Numerator.
$32 \div 8 = 4$ 4 replaces 32 in the Denominator.

$\frac{24}{32}$ **reduced to its Lowest Terms is** $\frac{3}{4}$

Example: Reduce $\frac{6}{9}$ to its Lowest Terms.

1. The GCF of 6 and 9 is 3.
2. $6 \div 3 = 2$ 2 replaces 6 in the Numerator.
$9 \div 3 = 3$ 3 replaces 9 in the Denominator.

$\frac{6}{9}$ **reduced to its Lowest Terms is** $\frac{2}{3}$

Example: Reduce $\frac{169}{52}$ to its Lowest Terms.

1. The GCF of 169 and 52 is 13.
2. $169 \div 13 = 13$ 13 replaces 169 in the Numerator.
$52 \div 13 = 4$ 4 replaces 52 in the Denominator.

$\frac{169}{52}$ **reduced to its Lowest Terms is** $\frac{13}{4}$

Example: Reduce $\frac{5}{7}$ to its Lowest Terms.

1. The GCF of 5 and 7 is 1, therefore $\frac{5}{7}$ is already in its Lowest Terms.

$\frac{5}{7}$ **is already in its Lowest Terms**

Reduce $\frac{16}{20}$ to its Lowest Terms.

1. The GCF of 16 and 20 is 4.

2. $16 \div 4 = 4$
4 replaces 16 in the Numerator.

$20 \div 4 = 5$
5 replaces 20 in the Denominator.

$\frac{16}{20}$ reduced to its Lowest Terms is $\frac{4}{5}$

1. Find the GCF of the Numerator and Denominator. The Numerator is 16 and the Denominator is 20.

2. Divide both the Numerator and Denominator by the GCF. The results of both of the divisions replace the Numerator and Denominator.

Reduce $\frac{5}{50}$ to its Lowest Terms.

1. The GCF of 5 and 50 is 5.

2. $5 \div 5 = 1$
1 replaces 5 in the Numerator.

$50 \div 5 = 10$
10 replaces 50 in the Denominator.

$\frac{5}{50}$ reduced to its Lowest Terms is $\frac{1}{10}$

1. Find the GCF of the Numerator and Denominator. The Numerator is 5 and the Denominator is 50.

2. Divide both the Numerator and Denominator by the GCF. The results of both of the divisions replace the Numerator and Denominator.

Reduce $\frac{56}{63}$ to its Lowest Terms.

1. The GCF of 56 and 63 is 7.

2. $56 \div 7 = 8$
8 replaces 56 in the Numerator.

$63 \div 7 = 9$
9 replaces 63 in the Denominator.

$\frac{56}{63}$ reduced to its Lowest Terms is $\frac{8}{9}$

1. Find the GCF of the Numerator and Denominator. The Numerator is 56 and the Denominator is 63.

2. Divide both the Numerator and Denominator by the GCF. The results of both of the divisions replace the Numerator and Denominator.

A fraction can be reduced to its Lowest Terms by being converted to a Whole Number if the Numerator is 0.

If the Numerator is 0, the Fraction can be converted to a Whole Number by dividing the Numerator by the Denominator. The result will always be 0.

Example: $\frac{0}{3}$

$0 \div 3 = 0$

$\frac{0}{3}$ **reduced to its Lowest Terms is 0**

Example: $\frac{0}{16}$

$0 \div 16 = 0$

$\frac{0}{16}$ **reduced to its Lowest Terms is 0**

Example: $\frac{0}{8}$

$0 \div 8 = 0$

$\frac{0}{8}$ **reduced to its Lowest Terms is 0**

Example: $\frac{0}{7}$

$0 \div 7 = 0$

$\frac{0}{7}$ **reduced to its Lowest Terms is 0**

There are several different methods that can be used to find the Least Common Multiple of 2 or more numbers. There are 3 listed below.

Method 1
List the multiples of each number, and look for the smallest number that appears in each list.

Method 2
1. Find the Greatest Common Factor (GCF) of the numbers
2. Multiply the numbers together
3. Divide the product of the numbers by the Greatest Common Factor.

Method 3
Factor each of the numbers into primes. For each different prime number in either of the factorizations, follow these steps:

1. Count the number of times it appears in each of the factorizations.
2. Take the largest of these two counts.
3. Write down that prime number as many times as the count in step 2.

The following page contains another method that can be used to find the Least Common Multiple. In this book we will use that method for finding the LCM since we believe it is the fastest method for finding the LCM.

Method 4

Step 1. Multiply the largest number by the number 1.
Step 2. Divide the result of the multiplication by the smaller number(s).
A. If there is not a remainder, then the result of the previous multiplication **(Step 1)** is the LCM.
B. If there is a remainder, then the result of the previous multiplication is not the LCM. Repeat Steps 1 and 2 until you get the LCM.
Note: Each time you repeat Step 1 use the next highest number to multiply the largest number by. For example multiply the number by 2, then 3, and so on.

Example: Find the Least Common Multiple of 5 and 4.

1. 5 is the largest number, so it will be multiplied by 1.
5 x 1 = 5
2. 5 ÷ 4 = 1 R1
B. There is a remainder, so 5 is not the Least Common Multiple for 5 and 4. Steps 1 and 2 will be repeated. In Step 1 the largest number will be multiplied by 2.

1. 5 is the largest number, so it will be multiplied by 2.
5 x 2 = 10
2. 10 ÷ 4 = 2 R2
B. There is a remainder, so 10 is not the Least Common Multiple for 5 and 4. Steps 1 and 2 will be repeated. In Step 1 the largest number will be multiplied by 3.

1. 5 is the largest number, so it will be multiplied by 3.
5 x 3 = 15
2. 15 ÷ 4 = 3 R3
B. There is a remainder, so 15 is not the Least Common Multiple for 5 and 4. Steps 1 and 2 will be repeated. In Step 1 the largest number will be multiplied by 4.

1. 5 is the largest number, so it will be multiplied by 4.
5 x 4 = 20
2. 20 ÷ 4 = 5
A. There is no remainder, so 20 is the Least Common Multiple for 5 and 4.

20 is the Least Common Multiple of 5 and 4.

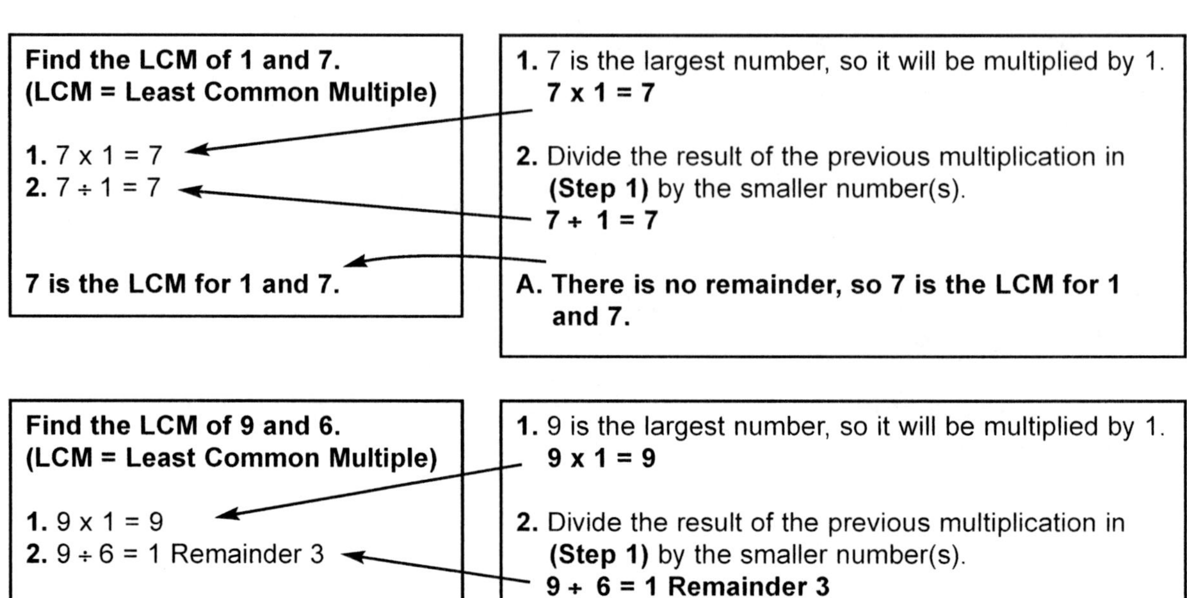

Find the LCM of 1 and 7.
(LCM = Least Common Multiple)

1. 7 x 1 = 7
2. 7 ÷ 1 = 7

7 is the LCM for 1 and 7.

1. 7 is the largest number, so it will be multiplied by 1.
7 x 1 = 7

2. Divide the result of the previous multiplication in **(Step 1)** by the smaller number(s).
7 ÷ 1 = 7

A. There is no remainder, so 7 is the LCM for 1 and 7.

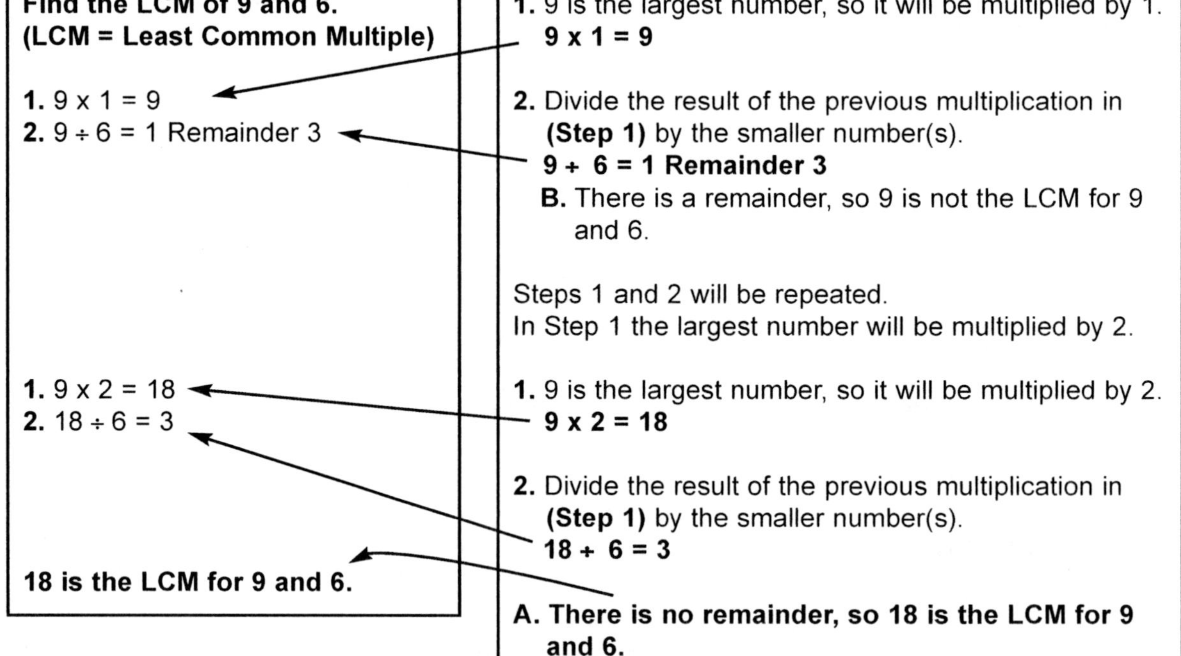

Find the LCM of 9 and 6.
(LCM = Least Common Multiple)

1. 9 x 1 = 9
2. 9 ÷ 6 = 1 Remainder 3

1. 9 x 2 = 18
2. 18 ÷ 6 = 3

18 is the LCM for 9 and 6.

1. 9 is the largest number, so it will be multiplied by 1.
9 x 1 = 9

2. Divide the result of the previous multiplication in **(Step 1)** by the smaller number(s).
9 ÷ 6 = 1 Remainder 3
B. There is a remainder, so 9 is not the LCM for 9 and 6.

Steps 1 and 2 will be repeated.
In Step 1 the largest number will be multiplied by 2.

1. 9 is the largest number, so it will be multiplied by 2.
9 x 2 = 18

2. Divide the result of the previous multiplication in **(Step 1)** by the smaller number(s).
18 ÷ 6 = 3

A. There is no remainder, so 18 is the LCM for 9 and 6.

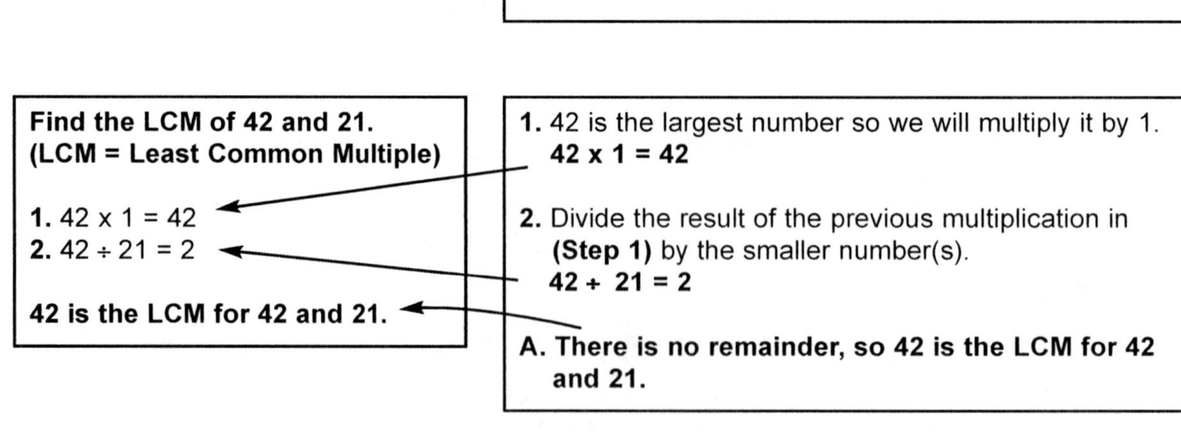

Find the LCM of 42 and 21.
(LCM = Least Common Multiple)

1. 42 x 1 = 42
2. 42 ÷ 21 = 2

42 is the LCM for 42 and 21.

1. 42 is the largest number so we will multiply it by 1.
42 x 1 = 42

2. Divide the result of the previous multiplication in **(Step 1)** by the smaller number(s).
42 ÷ 21 = 2

A. There is no remainder, so 42 is the LCM for 42 and 21.

The Math Mechanic Series: Fractions & Mixed Numbers Edition
Finding the Lowest Common Denominator - Pg. 39

The procedures for finding the Lowest Common Denominator are below:

Step 1. Find the Least Common Multiple of the Denominators in all fractions involved. ***This will become the Lowest Common Denominator.***

Step 2. Replacing the Numerator and Denominator.

- ***A.*** Divide the Lowest Common Denominator by the fraction's Denominator.
- ***B.*** Multiply the result of the previous division **(Step 2A)** by the Numerator. The result replaces the current Numerator of the fraction.
- ***C.*** Replace the fraction's Denominator with the Lowest Common Denominator.

Perform Step 2 on each fraction involved in the problem.

Example: Find the lowest common Denominator for $\frac{2}{3} + \frac{5}{7}$.

1. The Denominators in the fractions are 3 and 7.
The Least Common Multiple of 3 and 7 is 21, so the Lowest Common Denominator

for $\frac{2}{3} + \frac{5}{7}$ is 21.

$\mathbf{\frac{2}{3}}$

2A. 21 ÷ 3 = 7 (Dividing the Lowest Common Denominator by the fraction's Denominator)
2B. 7 x 2 = 14 (Multiplying the result from Step 2A by the fraction's Numerator)
14 replaces 2 as the Numerator.

2C. The Lowest Common Denominator replaces 3 as the Denominator.
21 replaces 3.

So $\frac{2}{3}$ becomes $\frac{14}{21}$.

$\mathbf{\frac{5}{7}}$

2A. 21 ÷ 7 = 3 (Dividing the Lowest Common Denominator by the fraction's Denominator)
2B. 3 x 5 = 15 (Multiplying the result from Step 2A by the fraction's Numerator)
15 replaces 5 as the Numerator.

2C. The Lowest Common Denominator replaces 7 as the Denominator.
21 replaces 7.

So $\frac{5}{7}$ becomes $\frac{15}{21}$.

$\mathbf{\frac{2}{3} + \frac{5}{7}}$ has become $\mathbf{\frac{14}{21} + \frac{15}{21}}$ after the Lowest Common Denominator has been found.

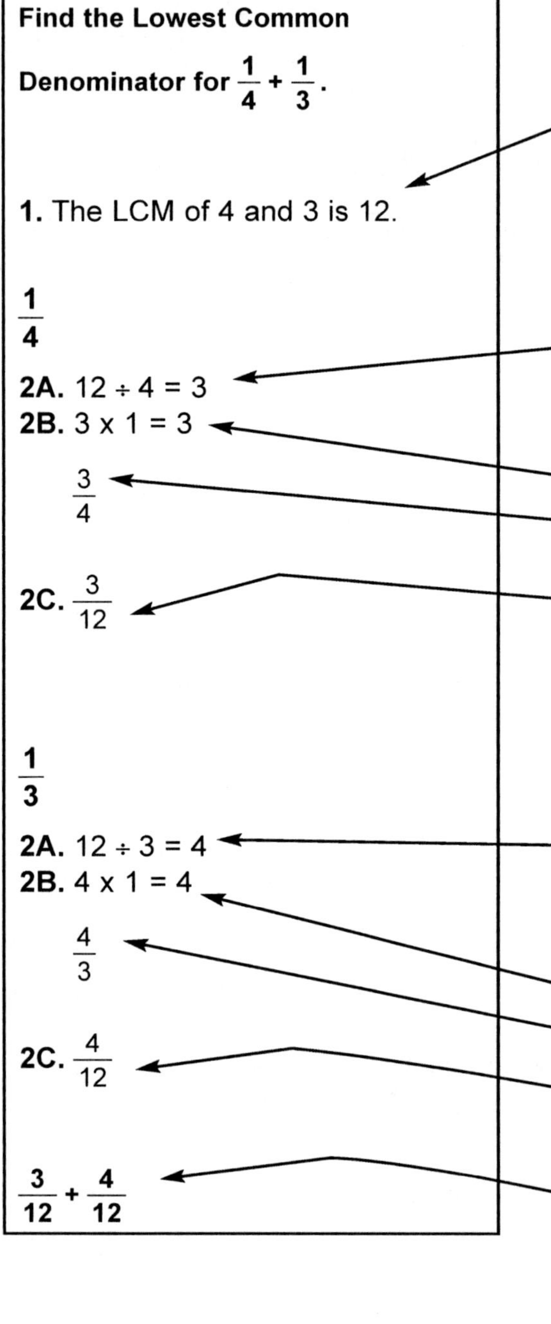

1. Find the LCM of the Denominators of all fractions involved.
The LCM of 4 and 3 is 12. The LCM becomes the Lowest Common Denominator.

2A. Divide the Lowest Common Denominator by the fraction's Denominator.

The Denominator of $\frac{1}{4}$ is 4.
$12 \div 4 = 3$

2B. Multiply the result of the previous division in (**Step 2A**) by the Numerator.

The Numerator of $\frac{1}{4}$ is 1.
$3 \times 1 = 3$
The result replaces the current Numerator of the fraction.

2C. Replace the fraction's Denominator with the Lowest Common Denominator.

2A. Divide the Lowest Common Denominator by the fraction's Denominator.

The Denominator of $\frac{1}{3}$ is 3.
$12 \div 3 = 4$

2B. Multiply the result of the previous division in (**Step 2A**) by the Numerator.

The Numerator of $\frac{1}{3}$ is 1.
$4 \times 1 = 4$
The result replaces the current Numerator of the fraction.

2C. Replace the fraction's Denominator with the Lowest Common Denominator.

$\frac{1}{4} + \frac{1}{3}$ becomes $\frac{3}{12} + \frac{4}{12}$ after finding the Lowest Common Denominator.

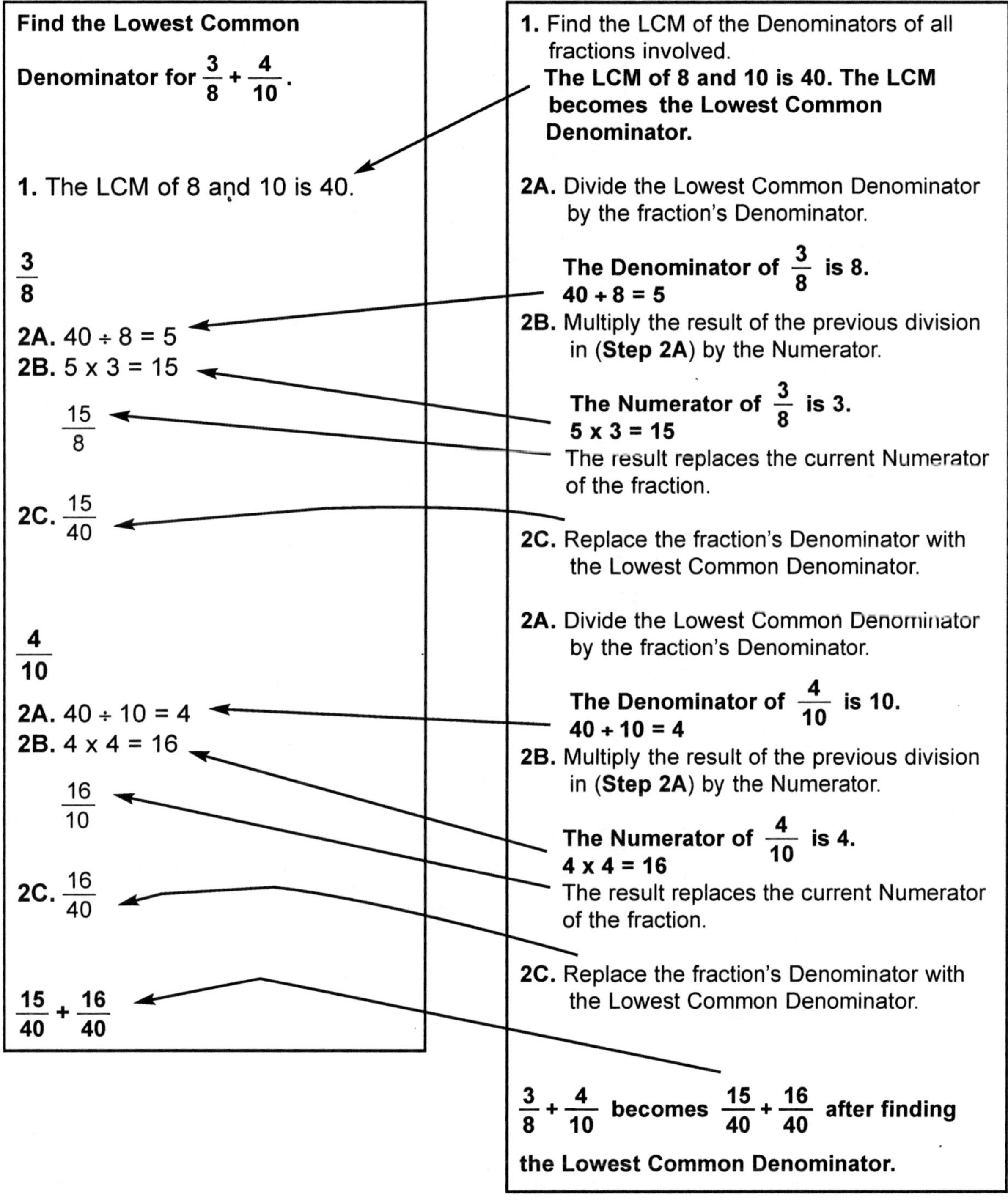

Find the Lowest Common Denominator for $\frac{3}{8} + \frac{4}{10}$.

1. The LCM of 8 and 10 is 40.

$\frac{3}{8}$

2A. 40 ÷ 8 = 5
2B. 5 x 3 = 15

$\frac{15}{8}$

2C. $\frac{15}{40}$

$\frac{4}{10}$

2A. 40 ÷ 10 = 4
2B. 4 x 4 = 16

$\frac{16}{10}$

2C. $\frac{16}{40}$

$\frac{15}{40} + \frac{16}{40}$

1. Find the LCM of the Denominators of all fractions involved.
The LCM of 8 and 10 is 40. The LCM becomes the Lowest Common Denominator.

2A. Divide the Lowest Common Denominator by the fraction's Denominator.

The Denominator of $\frac{3}{8}$ is 8.
40 ÷ 8 = 5

2B. Multiply the result of the previous division in (**Step 2A**) by the Numerator.

The Numerator of $\frac{3}{8}$ is 3.
5 x 3 = 15
The result replaces the current Numerator of the fraction.

2C. Replace the fraction's Denominator with the Lowest Common Denominator.

2A. Divide the Lowest Common Denominator by the fraction's Denominator.

The Denominator of $\frac{4}{10}$ is 10.
40 ÷ 10 = 4

2B. Multiply the result of the previous division in (**Step 2A**) by the Numerator.

The Numerator of $\frac{4}{10}$ is 4.
4 x 4 = 16
The result replaces the current Numerator of the fraction.

2C. Replace the fraction's Denominator with the Lowest Common Denominator.

$\frac{3}{8} + \frac{4}{10}$ becomes $\frac{15}{40} + \frac{16}{40}$ after finding the Lowest Common Denominator.

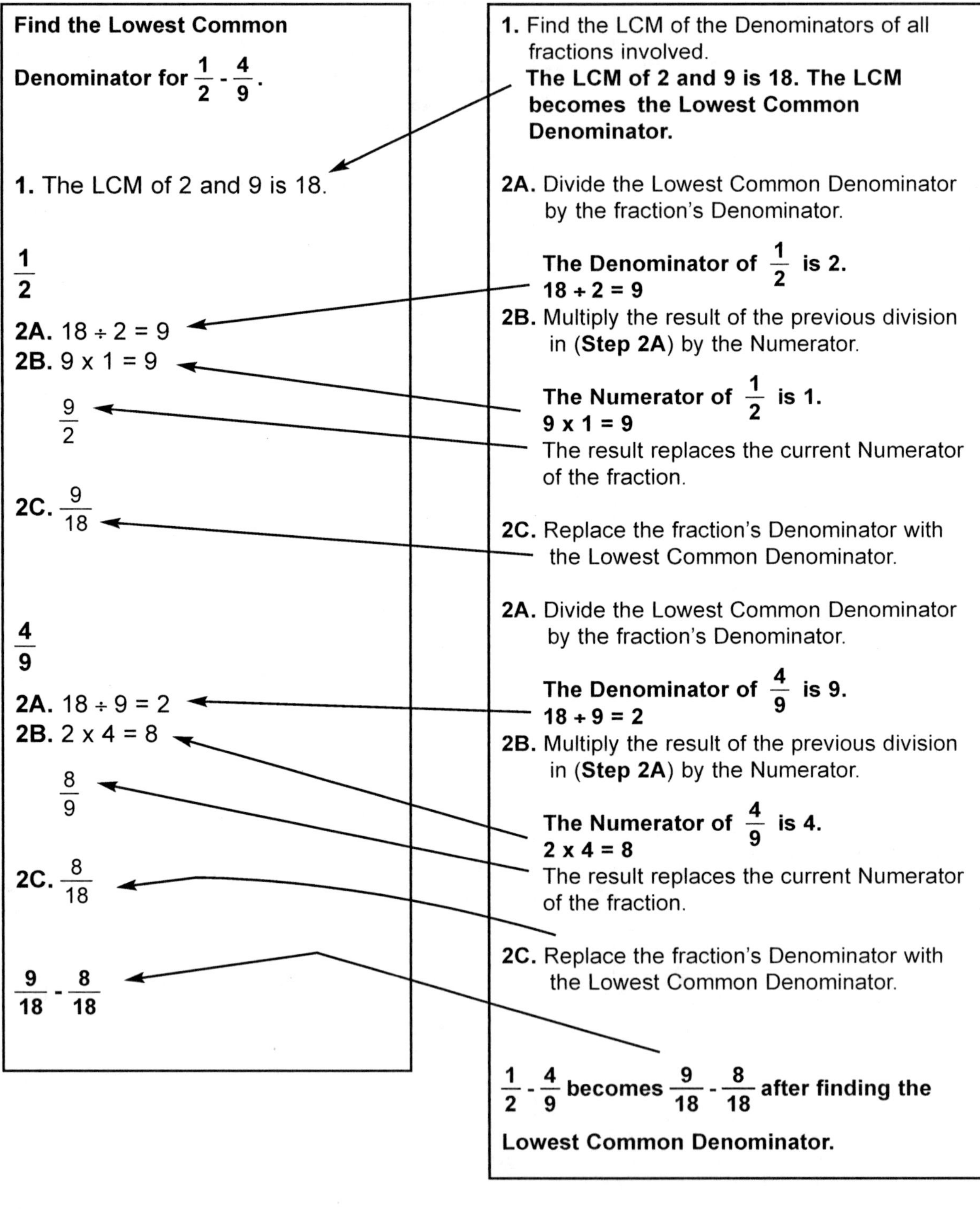
Find the Lowest Common Denominator for $\frac{1}{2} - \frac{4}{9}$.
1. The LCM of 2 and 9 is 18.
$\frac{1}{2}$
2A. 18 ÷ 2 = 9
2B. 9 x 1 = 9
$\frac{9}{2}$
2C. $\frac{9}{18}$
$\frac{4}{9}$
2A. 18 ÷ 9 = 2
2B. 2 x 4 = 8
$\frac{8}{9}$
2C. $\frac{8}{18}$
$\frac{9}{18} - \frac{8}{18}$
1. Find the LCM of the Denominators of all fractions involved.
The LCM of 2 and 9 is 18. The LCM becomes the Lowest Common Denominator.
2A. Divide the Lowest Common Denominator by the fraction's Denominator.
The Denominator of $\frac{1}{2}$ is 2.
18 ÷ 2 = 9
2B. Multiply the result of the previous division in (**Step 2A**) by the Numerator.
The Numerator of $\frac{1}{2}$ is 1.
9 x 1 = 9
The result replaces the current Numerator of the fraction.
2C. Replace the fraction's Denominator with the Lowest Common Denominator.
2A. Divide the Lowest Common Denominator by the fraction's Denominator.
The Denominator of $\frac{4}{9}$ is 9.
18 ÷ 9 = 2
2B. Multiply the result of the previous division in (**Step 2A**) by the Numerator.
The Numerator of $\frac{4}{9}$ is 4.
2 x 4 = 8
The result replaces the current Numerator of the fraction.
2C. Replace the fraction's Denominator with the Lowest Common Denominator.
$\frac{1}{2} - \frac{4}{9}$ becomes $\frac{9}{18} - \frac{8}{18}$ after finding the Lowest Common Denominator.

The procedures for adding fractions with the same Denominator are below:

If the fractions have the same Denominator

Step 1. Add the Numerators together. The result will become the answer's Numerator. The answer's Denominator will be the same as the Denominator in the fractions that were added.

Step 2. If the answer is an Improper Fraction, change it to a Mixed Number or a Whole Number.

Step 3. If there is a fraction in the answer, reduce it to its Lowest Terms if possible.

Example: $\frac{3}{4} + \frac{2}{4}$

Step 1. Add the Numerators together.

$\mathbf{3 + 2 = 5}$

$\frac{3}{4} + \frac{2}{4} = \frac{5}{4}$

Step 2. If the answer is an Improper Fraction, change it to a Mixed Number or a Whole Number.

$\frac{5}{4}$ is an Improper Fraction, so it will be changed to a Mixed Number.

$\frac{5}{4}$ changes to $1\frac{1}{4}$.

Step 3. If there is a fraction in the answer, reduce it to its Lowest Terms if possible.

$\frac{1}{4}$ is already in its Lowest Terms.

The answer to this problem is $1\frac{1}{4}$.

Step-by-Step Examples of Adding Fractions with the Same Denominator - Pg. 44

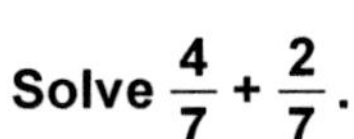

Solve $\frac{4}{7} + \frac{2}{7}$.

1. 4 + 2 = 6

$\frac{4}{7} + \frac{2}{7} = \frac{6}{7}$

$\frac{6}{7}$

1. Add the Numerators together. The Numerators are 4 and 2. The result will become the answer's Numerator.

4 + 2 = 6.

The answer's Denominator will be the same as the Denominator in the fractions that were added.

The Denominator in both fractions that were added is 7, so 7 will be the Denominator in the answer.

2. If the answer is an Improper Fraction, change it to a Mixed Number or a Whole Number.

$\frac{6}{7}$ **is not an Improper Fraction.**

3. If there is a fraction in the answer, reduce it to its Lowest Terms if possible.

$\frac{6}{7}$ **is already in its Lowest Terms.**

$\frac{6}{7}$ **is the answer to this problem.**

Solve $\frac{5}{18} + \frac{5}{18}$.

1. 5 + 5 = 10

$\frac{5}{18} + \frac{5}{18} = \frac{10}{18}$

3. $\frac{10}{18} = \frac{5}{9}$

$\frac{5}{9}$

1. Add the Numerators together. The Numerators of both fractions is 5. The result will become the answer's Numerator.

5 + 5 = 10.

The answer's Denominator will be the same as the Denominator in the fractions that were added.

The Denominator in both fractions that were added is 18, so 18 will be the Denominator in the answer.

2. If the answer is an Improper Fraction, change it to a Mixed Number or a Whole Number.

$\frac{10}{18}$ **is not an Improper Fraction.**

3. If there is a fraction in the answer, reduce it to its Lowest Terms if possible.

$\frac{10}{18}$ **reduced to its Lowest Terms is** $\frac{5}{9}$.

$\frac{5}{9}$ **is the answer to this problem.**

The Math Mechanic Series: Fractions & Mixed Numbers Edition
Step-by-Step Examples of Adding Fractions with the Same Denominator - Pg. 45

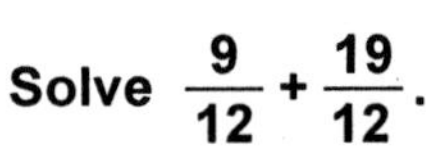

Solve $\frac{9}{12} + \frac{19}{12}$.

1. $9 + 19 = 28$

$\frac{9}{12} + \frac{19}{12} = \frac{28}{12}$

2. $\frac{28}{12} = 2\frac{4}{12}$

3. $2\frac{4}{12} = 2\frac{1}{3}$

$2\frac{1}{3}$

1. Add the Numerators together. The Numerators are 9 and 19. The result will become the answer's Numerator.

9 + 19 = 28.

The answer's Denominator will be the same as the Denominator in the fractions that were added.

The Denominator in both fractions that were added is 12, so 12 will be the Denominator in the answer.

2. If the answer is an Improper Fraction, change it to a Mixed Number or a Whole Number.

$\frac{28}{12}$ is an Improper Fraction, so it will be changed to a Mixed Number.

$\frac{28}{12} = 2\frac{4}{12}$

3. If there is a fraction in the answer, reduce it to its Lowest Terms if possible.

$\frac{4}{12}$ reduced to its Lowest Terms is $\frac{1}{3}$.

$2\frac{1}{3}$ is the answer to this problem.

Solve $\frac{5}{7} + \frac{8}{7}$.

1. $5 + 8 = 13$

$\frac{5}{7} + \frac{8}{7} = \frac{13}{7}$

2. $\frac{13}{7} = 1\frac{6}{7}$

$1\frac{6}{7}$

1. Add the Numerators together. The Numerators are 5 and 8. The result will become the answer's Numerator.

5 + 8 = 13.

The answer's Denominator will be the same as the Denominator in the fractions that were added.

The Denominator in both fractions that were added is 7, so 7 will be the Denominator in the answer.

2. If the answer is an Improper Fraction, change it to a Mixed Number or a Whole Number.

$\frac{13}{7}$ is an Improper Fraction, so it will be changed to a Mixed Number.

$\frac{13}{7} = 1\frac{6}{7}$

3. If there is a fraction in the answer, reduce it to its Lowest Terms if possible.

$\frac{6}{7}$ is already in its Lowest Terms.

$1\frac{6}{7}$ is the answer to this problem.

The procedures for adding fractions with different Denominators are below:

If the fractions do not all have the same Denominator

Step 1. Change the fractions so that they have the same Denominator by finding the Lowest Common Denominator of all fractions involved.

Step 2. Add the Numerators together. The result will become the answer's Numerator. The answer's Denominator will be the same as the Denominator in the fractions that were added.

Step 3. If the answer is an Improper Fraction, change it to a Mixed Number or a Whole Number.

Step 4. If there is a fraction in the answer, reduce it to its Lowest Terms if possible.

Example: $\frac{4}{9} + \frac{3}{5}$

Step 1. The Lowest Common Denominator of both fractions is 45. This changes both fractions.

$\frac{4}{9} = \frac{20}{45}$ and $\frac{3}{5} = \frac{27}{45}$

Step 2. Add the Numerators together.

20 + 27 = 47

$$\begin{array}{rl} & \frac{4}{9} = \frac{20}{45} \\ + & \frac{3}{5} = \frac{27}{45} \\ \hline & \quad = \frac{47}{45} \end{array}$$

Step 3. If the answer is an Improper Fraction, change it to a Mixed Number or a Whole Number.

$\frac{47}{45}$ is an Improper Fraction, so it will be changed to a Mixed Number.

$\frac{47}{45}$ changes to $1\frac{2}{45}$.

Step 4. If there is a fraction in the answer, reduce it to its Lowest Terms if possible.

$\frac{2}{45}$ is already in its Lowest Terms.

The answer to this problem is $1\frac{2}{45}$.

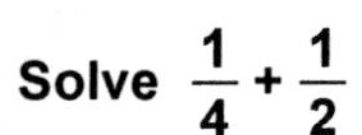

Solve $\frac{1}{4} + \frac{1}{2}$.

1. $\frac{1}{4} = \frac{1}{4}$ and $\frac{1}{2} = \frac{2}{4}$

2.

$$\begin{array}{r} \frac{1}{4} = \frac{1}{4} \\ + \ \frac{1}{2} = \frac{2}{4} \\ \hline = \frac{3}{4} \end{array}$$

$\frac{3}{4}$

1. Change the fractions so that they have the same Denominator by finding the Lowest Common Denominator of all fractions involved.

$\frac{1}{4}$ stays at $\frac{1}{4}$ and $\frac{1}{2}$ changes to $\frac{2}{4}$

2. Add the Numerators together. The Numerators are 1 and 2. The result will become the answer's Numerator.

1 + 2 = 3

The answer's Denominator will be the same as the Denominator in the fractions that were added.

The Denominator in both fractions that were added is 4, so 4 will be the Denominator in the answer.

3. If the answer is an Improper Fraction, change it to a Mixed Number or a Whole Number.

$\frac{3}{4}$ is not an Improper Fraction.

4. If there is a fraction in the answer, reduce it to its Lowest Terms if possible.

$\frac{3}{4}$ is already in its Lowest Terms.

$\frac{3}{4}$ is the answer to this problem.

Solve $\frac{4}{5} + \frac{2}{3}$.

1. $\frac{4}{5} = \frac{12}{15}$ and $\frac{2}{3} = \frac{10}{15}$

2.

$$\begin{array}{r} \frac{4}{5} = \frac{12}{15} \\ + \frac{2}{3} = \frac{10}{15} \\ \hline = \frac{22}{15} \end{array}$$

3. $\frac{22}{15} = 1\frac{7}{15}$

$\mathbf{1\frac{7}{15}}$

1. Change the fractions so that they have the same Denominator by finding the Lowest Common Denominator of all fractions involved.

$\frac{4}{5}$ **changes to** $\frac{12}{15}$ **and** $\frac{2}{3}$ **changes to** $\frac{10}{15}$ **.**

2. Add the Numerators together. The Numerators are 12 and 10. The result will become the answer's Numerator.

12 + 10 = 22.

The answer's Denominator will be the same as the Denominator in the fractions that were added.

The Denominator in both fractions that were added is 15, so 15 will be the Denominator in the answer.

3. If the answer is an Improper Fraction, change it to a Mixed Number or a Whole Number.

$\frac{22}{15}$ **is an Improper Fraction, so it will be changed to a Mixed Number.**

$\frac{22}{15} = 1\frac{7}{15}$

4. If there is a fraction in the answer, reduce it to its Lowest Terms if possible.

$\frac{7}{15}$ **is already in its Lowest Terms.**

$1\frac{7}{15}$ **is the answer to this problem.**

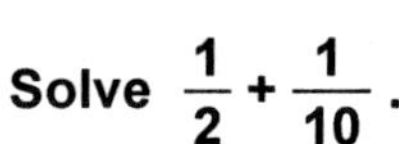

Solve $\frac{1}{2} + \frac{1}{10}$.

1. $\frac{1}{2} = \frac{5}{10}$ and $\frac{1}{10} = \frac{1}{10}$

2.
$$\begin{array}{r} \frac{1}{2} = \frac{5}{10} \\ + \; \frac{1}{10} = \frac{1}{10} \\ \hline = \frac{6}{10} \end{array}$$

4. $\frac{6}{10} = \frac{3}{5}$

$\frac{3}{5}$

1. Change the fractions so that they have the same Denominator by finding the Lowest Common Denominator of all fractions involved.

$\frac{1}{2}$ changes to $\frac{5}{10}$ and $\frac{1}{10}$ stays at $\frac{1}{10}$.

2. Add the Numerators together. The Numerators are 5 and 1. The result will become the answer's Numerator.

5 + 1 = 6.

The answer's Denominator will be the same as the Denominator in the fractions that were added.

The Denominator in both fractions that were added is 10, so 10 will be the Denominator in the answer.

3. If the answer is an Improper Fraction, change it to a Mixed Number or a Whole Number.

$\frac{6}{10}$ is not an Improper Fraction.

4. If there is a fraction in the answer, reduce it to its Lowest Terms if possible.

$\frac{6}{10}$ reduced to its Lowest Terms is $\frac{3}{5}$.

$\frac{3}{5}$ is the answer to this problem.

Solve $\frac{4}{15} + \frac{5}{6}$.

1. $\frac{4}{15} = \frac{8}{30}$ and $\frac{5}{6} = \frac{25}{30}$

2. $\frac{4}{15} = \frac{8}{30}$

$+ \frac{5}{6} = \frac{25}{30}$

$= \frac{33}{30}$

3. $\frac{33}{30} = 1\frac{3}{30}$

4. $1\frac{3}{30} = 1\frac{1}{10}$

$\mathbf{1\frac{1}{10}}$

1. Change the fractions so that they have the same Denominator by finding the Lowest Common Denominator of all fractions involved.

$\frac{4}{15}$ changes to $\frac{8}{30}$ and $\frac{5}{6}$ changes to $\frac{25}{30}$.

2. Add the Numerators together. The Numerators are 8 and 25. The result will become the answer's Numerator.

8 + 25 = 33.

The answer's Denominator will be the same as the Denominator in the fractions that were added.

The Denominator in both fractions that were added is 30, so 30 will be the Denominator in the answer.

3. If the answer is an Improper Fraction, change it to a Mixed Number or a Whole Number.

$\frac{33}{30}$ is an Improper Fraction, so it will be changed to a Mixed Number.

$\frac{33}{30} = 1\frac{3}{30}$

4. If there is a fraction in the answer, reduce it to its Lowest Terms if possible.

$\frac{3}{30}$ reduced to its Lowest Terms is $\frac{1}{10}$.

$1\frac{1}{10}$ is the answer to this problem.

The procedures for adding Mixed Numbers with the same Denominator are below:

Step 1. Add the Whole Numbers together.
Step 2. Add the Numerators together. The result will become the answer's Numerator. The answer's Denominator will be the same as the Denominator in the fractions that were added.
Step 3. Reduce the answer's fraction part to its Lowest Terms if possible.
Note: You can add the Numerators together before adding the Whole Numbers together if you want.

Example: $8\frac{4}{7} + 6\frac{2}{7}$

Step 1. Add the Whole Numbers together.

$8\frac{4}{7} + 6\frac{2}{7}$

8 + 6 = 14

Step 2. Add the Numerators together.

$8\frac{4}{7} + 6\frac{2}{7}$

4 + 2 = 6

$8\frac{4}{7} + 6\frac{2}{7} = 14\frac{6}{7}$

Step 3. Reduce the answer's fraction to its Lowest Terms if possible.

The fraction part of $14\frac{6}{7}$ is $\frac{6}{7}$, this fraction is already in its Lowest Terms.

The answer to this problem is $14\frac{6}{7}$.

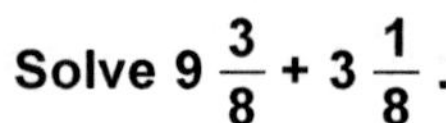

Solve $9\frac{3}{8} + 3\frac{1}{8}$.

1. $9 + 3 = 12$

2. $3 + 1 = 4$

$9\frac{3}{8} + 3\frac{1}{8} = 12\frac{4}{8}$

3. $12\frac{4}{8} = 12\frac{1}{2}$

$12\frac{1}{2}$

1. Add the Whole Numbers together.

$9 + 3 = 12$.

2. Add the Numerators together. The Numerators are 3 and 1. The result will become the answer's Numerator.

$3 + 1 = 4$

The answer's Denominator will be the same as the Denominator in the fractions that were added.

The Denominator in both fractions that were added is 8, so 8 will be the Denominator in the answer.

3. Reduce the answer's fraction part to its Lowest Terms if possible.

The fraction part is $\frac{4}{8}$.

$\frac{4}{8}$ reduced to its Lowest Terms is $\frac{1}{2}$.

$12\frac{1}{2}$ is the answer to this problem.

Solve $10\frac{7}{13} + 6\frac{3}{13}$.

1. 10 + 6 = 16

2. 7 + 3 = 10

$10\frac{7}{13} + 6\frac{3}{13} = 16\frac{10}{13}$

$16\frac{10}{13}$

1. Add the Whole Numbers together.

10 + 6 = 16.

2. Add the Numerators together. The Numerators are 7 and 3. The result will become the answer's Numerator.

7 + 3 = 10.

The answer's Denominator will be the same as the Denominator in the fractions that were added.

The Denominator in both fractions that were added is 13, so 13 will be the Denominator in the answer.

3. Reduce the answer's fraction part to its Lowest Terms if possible.

The fraction part is $\frac{10}{13}$.

$\frac{10}{13}$ is already in its Lowest Terms.

$16\frac{10}{13}$ is the answer to this problem.

The procedures for adding Mixed Numbers with different Denominators are below:

Step 1. Change the fractions so that they have the same Denominator by finding the Lowest Common Denominator of all fractions involved.

Step 2. Add the Whole Numbers together.

Step 3. Add the Numerators together. The result will become the answer's Numerator. The answer's Denominator will be the same as the Denominator in the fractions that were added.

Step 4. Reduce the answer's fraction part to its Lowest Terms if possible.

Note: You can add the Numerators together before adding the Whole Numbers together if you want.

Example: $10\frac{8}{15} + 9\frac{1}{60}$

Step 1. The Lowest Common Denominator of both fractions is 60.

This will change $\frac{8}{15}$ to $\frac{32}{60}$ and $\frac{1}{60}$ will stay $\frac{1}{60}$.

$\frac{8}{15} = \frac{32}{60}$ and $\frac{1}{60} = \frac{1}{60}$

Step 2. Add the Whole Numbers together.

10 + 9 = 19

$$\begin{array}{rcr} 10\frac{8}{15} & = & 10\frac{32}{60} \\ +\ 9\frac{1}{60} & = & 9\frac{1}{60} \\ \hline & = & 19 \end{array}$$

Step 3. Add the Numerators together.

32 + 1 = 33

$$\begin{array}{rcr} 10\frac{8}{15} & = & 10\frac{32}{60} \\ +\ 9\frac{1}{60} & = & 9\frac{1}{60} \\ \hline & = & 19\frac{33}{60} \end{array}$$

Step 4. Reduce the answer's fraction part to its Lowest Terms if possible.

The fraction part of $19\frac{33}{60}$ is $\frac{33}{60}$.

The GCF of the Numerator 33 and the Denominator 60 is 3.

$33 \div 3 = 11$

$60 \div 3 = 20$

The Lowest Terms of $\frac{33}{60}$ is $\frac{11}{20}$.

The answer to this problem is $19\frac{11}{20}$.

Solve $8\frac{5}{6} + 4\frac{1}{8}$.

1. $8\frac{5}{6} = 8\frac{20}{24}$ and $4\frac{1}{8} = 4\frac{3}{24}$

2. $8 + 4 = 12$

3. $20 + 3 = 23$

$$\begin{array}{r} 8\frac{5}{6} = 8\frac{20}{24} \\ +\ 4\frac{1}{8} = 4\frac{3}{24} \\ \hline = 12\frac{23}{24} \end{array}$$

$12\frac{23}{24}$

1. Change the fractions so that they have the same Denominator by finding the Lowest Common Denominator of all fractions involved.

$\frac{5}{6}$ changes to $\frac{20}{24}$ and $\frac{1}{8}$ changes to $\frac{3}{24}$.

2. Add the Whole Numbers together.

$8 + 4 = 12$.

3. Add the Numerators together. The Numerators are 20 and 3. The result will become the answer's Numerator.

$20 + 3 = 23$.

The answer's Denominator will be the same as the Denominator in the fractions that were added.

The Denominator in both fractions that were added is 24, so 24 will be the Denominator in the answer.

4. Reduce the answer's fraction part to its Lowest Terms if possible.

The fraction part is $\frac{23}{24}$.

$\frac{23}{24}$ is already in its Lowest Terms.

$12\frac{23}{24}$ is the answer to this problem.

Solve $7\frac{1}{12} + 1\frac{4}{15}$.

1. $7\frac{1}{12} = 7\frac{5}{60}$ and $1\frac{4}{15} = 1\frac{16}{60}$

2. 7 + 1 = 8

3. 5 + 16 = 21

$$7\frac{1}{12} = 7\frac{5}{60}$$
$$+\ 1\frac{4}{15} = 1\frac{16}{60}$$
$$= 8\frac{21}{60}$$

4. $8\frac{21}{60} = 8\frac{7}{20}$

$8\frac{7}{20}$

1. Change the fractions so that they have the same Denominator by finding the Lowest Common Denominator of all fractions involved.

$\frac{1}{12}$ changes to $\frac{5}{60}$ and $\frac{4}{15}$ changes to $\frac{16}{60}$.

2. Add the Whole Numbers together.

7 + 1 = 8

3. Add the Numerators together. The Numerators are 5 and 16. The result will become the answer's Numerator.

5 + 16 = 21

The answer's Denominator will be the same as the Denominator in the fractions that were added.

The Denominator in both fractions that were added is 60, so 60 will be the Denominator in the answer.

4. Reduce the answer's fraction part to its Lowest Terms if possible.

The fraction part is $\frac{21}{60}$.

$\frac{21}{60}$ reduced to its Lowest Terms is $\frac{7}{20}$.

$8\frac{7}{20}$ is the answer to this problem.

The procedures for subtracting fractions with the same Denominator are below:

Step 1. Subtract the second Numerator from the first Numerator. The result will become the answer's Numerator. The answer's Denominator will be the same as the Denominator in the fractions that were in the subtraction problem.
Step 2. If the answer is an Improper Fraction, change it to a Mixed Number or a Whole Number.
Step 3. If there is a fraction in the answer, reduce it to its Lowest Terms if possible.

Example: $\frac{7}{17} - \frac{6}{17}$

Step 1. Subtract the second Numerator from the first.
The second Numerator is 6 and the first Numerator is 7.

7 – 6 = 1

$$\frac{7}{17} - \frac{6}{17} = \frac{1}{17}$$

Step 2. If the answer is an Improper Fraction, change it to a Mixed Number or a Whole Number.

$\frac{1}{17}$ is not an Improper Fraction.

Step 3. If there is a fraction in the answer, reduce it to its Lowest Terms if possible.

The fraction $\frac{1}{17}$ is already in its Lowest Terms.

The answer to this problem is $\frac{1}{17}$.

Solve $\frac{6}{7} - \frac{3}{7}$.

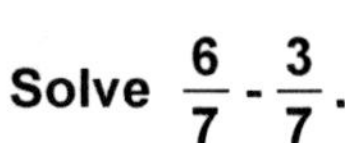

1. 6 - 3 = 3

$\frac{6}{7} - \frac{3}{7} = \frac{3}{7}$

$\frac{3}{7}$

1. Subtract the second Numerator from the first Numerator. The result will become the answer's Numerator.

6 - 3 = 3

The answer's Denominator will be the same as the Denominator in the fractions that were in the subtraction problem.

The Denominator in both fractions that were in the subtraction problem is 7, so 7 will be the Denominator in the answer.

2. If the answer is an Improper Fraction, change it to a Mixed Number or a Whole Number.

$\frac{3}{7}$ **is not an Improper Fraction.**

3. If there is a fraction in the answer, reduce it to its Lowest Terms if possible.

$\frac{3}{7}$ **is already in its Lowest Terms.**

$\frac{3}{7}$ **is the answer to this problem.**

Solve $\frac{19}{25} - \frac{4}{25}$.

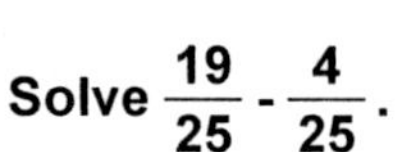

1. 19 - 4 = 15

$\frac{19}{25} - \frac{4}{25} = \frac{15}{25}$

3. $\frac{15}{25} = \frac{3}{5}$

$\frac{3}{5}$

1. Subtract the second Numerator from the first Numerator. The result will become the answer's Numerator.

19 - 4 = 15

The answer's Denominator will be the same as the Denominator in the fractions that were in the subtraction problem.

The Denominator in both fractions that were in the subtraction problem is 25, so 25 will be the Denominator in the answer.

2. If the answer is an Improper Fraction, change it to a Mixed Number or a Whole Number.

$\frac{15}{25}$ **is not an Improper Fraction.**

3. If there is a fraction in the answer, reduce it to its Lowest Terms if possible.

$\frac{15}{25}$ **reduced to its Lowest Terms is** $\frac{3}{5}$.

$\frac{3}{5}$ **is the answer to this problem.**

The procedures for subtracting fractions with different Denominators are below:

Step 1. Change the fractions so that they have the same Denominator by finding the Lowest Common Denominator of all fractions involved.
Step 2. Subtract the second Numerator from the first Numerator. The result will become the answer's Numerator. The answer's Denominator will be the same as the Denominator in the fractions that were used for the actual subtraction.
Step 3. If the answer is an Improper Fraction, change it to a Mixed Number or a Whole Number.
Step 4. If there is a fraction in the answer, reduce it to its Lowest Terms if possible.

Example: $\frac{3}{4} - \frac{10}{18}$

Step 1. The Lowest Common Denominator of both fractions is 36. This changes both fractions.

$\frac{3}{4} = \frac{27}{36}$ and $\frac{10}{18} = \frac{20}{36}$

Step 2. Subtract the second Numerator from the first Numerator. The second Numerator is 20 and the first Numerator is 27.

27 – 20 = 7

$$\begin{array}{r} \frac{3}{4} = \frac{27}{36} \\ - \ \frac{10}{18} = \frac{20}{36} \\ \hline = \frac{7}{36} \end{array}$$

Step 3. If the answer is an Improper Fraction, change it to a Mixed Number or a Whole Number.

$\frac{7}{36}$ is not an Improper fraction.

Step 4. If there is a fraction in the answer, reduce it to its Lowest Terms if possible.

The fraction $\frac{7}{36}$ is already in its Lowest Terms.

The answer to this problem is $\frac{7}{36}$.

Solve $\frac{9}{10} - \frac{4}{15}$.

1. $\frac{9}{10} = \frac{27}{30}$ and $\frac{4}{15} = \frac{8}{30}$

$\frac{9}{10} - \frac{4}{15} = \frac{27}{30} - \frac{8}{30}$

2. 27 - 8 = 19

$\frac{27}{30} - \frac{8}{30} = \frac{19}{30}$

$\mathbf{\frac{19}{30}}$

1. Change the fractions so that they have the same Denominator by finding the Lowest Common Denominator of all fractions involved.

$\frac{9}{10}$ **changes to** $\frac{27}{30}$ **and** $\frac{4}{15}$ **changes to** $\frac{8}{30}$

2. Subtract the second Numerator from the first Numerator. The result will become the answer's Numerator. The first Numerator is 27 and the second Numerator is 8.

27 - 8 = 19

The answer's Denominator will be the same as the Denominator in the fractions that were used for the actual subtraction.

The Denominator in both fractions that were used for the actual subtraction is 30, so 30 will be the Denominator in the answer.

3. If the answer is an Improper Fraction, change it to a Mixed Number or a Whole Number.

$\frac{19}{30}$ **is not an Improper Fraction.**

4. If there is a fraction in the answer, reduce it to its Lowest Terms if possible.

$\frac{19}{30}$ **is already in its Lowest Terms.**

$\frac{19}{30}$ **is the answer to this problem.**

Solve $\frac{11}{14} - \frac{15}{28}$.

1. $\frac{11}{14} = \frac{22}{28}$ and $\frac{15}{28} = \frac{15}{28}$

$\frac{11}{14} - \frac{15}{28} = \frac{22}{28} - \frac{15}{28}$

2. 22 - 15 = 7

$\frac{22}{28} - \frac{15}{28} = \frac{7}{28}$

4. $\frac{7}{28} = \frac{1}{4}$

$\frac{1}{4}$

1. Change the fractions so that they have the same Denominator by finding the Lowest Common Denominator of all fractions involved.

$\frac{11}{14}$ changes to $\frac{22}{28}$ and $\frac{15}{28}$ stays at $\frac{15}{28}$

2. Subtract the second Numerator from the first Numerator. The result will become the answer's Numerator. The first Numerator is 22 and the second Numerator is 15.

22 - 15 = 7

The answer's Denominator will be the same as the Denominator in the fractions that were used for the actual subtraction.

The Denominator in both fractions that were used for the actual subtraction is 28, so 28 will be the Denominator in the answer.

3. If the answer is an Improper Fraction, change it to a Mixed Number or a Whole Number.

$\frac{7}{28}$ is not an Improper Fraction.

4. If there is a fraction in the answer, reduce it to its Lowest Terms if possible.

$\frac{7}{28}$ reduced to its Lowest Terms is $\frac{1}{4}$.

$\frac{1}{4}$ is the answer to this problem.

The Math Mechanic Series: Fractions & Mixed Numbers Edition
Subtracting Mixed Numbers and Borrowing - Pg. 62

Some subtraction problems involving Mixed Numbers will require you to Borrow. Borrowing in the case of subtracting Mixed Numbers, means that you will borrow from the Whole Number part of the **Minuend** and add to its Fraction part. The **Minuend** is the value to be subtracted from and the **Subtrahend** is the value that will be used to subtract.

When will you need to Borrow?
Answer: *You will need to Borrow when after you have found the Lowest Common Denominator, the Numerator in the Subtrahend is larger than the Numerator in the Minuend.*

The procedures for Borrowing are below:
Step 1: Subtract 1 from the Whole Number part of the Minuend.
Step 2: Add the 1 that was borrowed from the Whole Number part of the Minuend to the Fraction part of the Minuend. The Numerator will be changed by this action.
This is done by creating a fraction in the form of Denominator over the Denominator.
Note: Use the Denominator in the Fraction part of the Minuend.

Example: $\frac{1}{4} + \frac{4}{4}$ The Denominator in $\frac{1}{4}$ is 4, so we added $\frac{4}{4}$ to $\frac{1}{4}$.

Example: $6\frac{3}{7} - 3\frac{5}{7}$

In the example above, the Lowest Common Denominator has already been found. The Numerator of the Subtrahend, which is 5, is larger than the Numerator of the Minuend, which is 3.

Step 1. Subtract 1 from the Whole Number part of the Minuend.
$6 - 1 = 5$
Step 2. Add the 1 that was borrowed in Step 1 to the Fraction part of the Minuend. This is done in the form of Denominator over Denominator. The Numerator will be changed by this action.

The Denominator of the Minuend is 7, so you will add $\frac{7}{7}$ to $\frac{3}{7}$.

$\frac{3}{7} + \frac{7}{7} = \frac{10}{7}$

The Minuend now looks like this: $5\frac{10}{7}$.

The problem now becomes: $5\frac{10}{7} - 3\frac{5}{7}$.

Example: $9\frac{4}{5} - 2\frac{6}{5}$

In the example above, the Lowest Common Denominator has already been found. The Numerator of the Subtrahend, which is 6, is larger than the Numerator of the Minuend, which is 4.

Step 1. Subtract 1 from the Whole Number part of the Minuend.
$9 - 1 = 8$
Step 2. Add the 1 that was borrowed in Step 1 to the Fraction part of the Minuend. This is done in the form of Denominator over Denominator. The Numerator will be changed by this action.

The Denominator of the Minuend is 5, so you will add $\frac{5}{5}$ to $\frac{4}{5}$.

$\frac{4}{5} + \frac{5}{5} = \frac{9}{5}$

The Minuend now looks like this: $8\frac{9}{5}$.

The problem now becomes: $8\frac{9}{5} - 2\frac{6}{5}$.

The procedures for subtracting Mixed Numbers with the same Denominator are below:

Step 1. Borrow if necessary.
Step 2. Subtract the second Whole Number from the first Whole Number.
Step 3. Subtract the second Numerator from the first Numerator.
Step 4. If there is a fraction in the answer, reduce it to its Lowest Terms if possible.
Note: You can subtract the Numerators before subtracting the Whole Numbers if you want.

Example: $6\frac{8}{9} - 1\frac{2}{9}$

Step 1. The Numerator of the Subtrahend is not larger than the Numerator of the Minuend, therefore it is not necessary to borrow.

Step 2. Subtract the second Whole Number from the first Whole Number.

$6\frac{8}{9} - 1\frac{2}{9}$

6 – 1 = 5

Step 3. Subtract the second Numerator from the first Numerator.

$6\frac{8}{9} - 1\frac{2}{9}$

8 – 2 = 6

$6\frac{8}{9} - 1\frac{2}{9} = 5\frac{6}{9}$

Step 4. Reduce the answer's fraction part to its Lowest Terms if possible.

The fraction part of $5\frac{6}{9}$ is $\frac{6}{9}$.

The GCF of the Numerator 6 and the Denominator 9 is 3.

$6 \div 3 = 2$
$9 \div 3 = 3$

The Lowest Terms of $\frac{6}{9}$ is $\frac{2}{3}$.

The answer to this problem is $5\frac{2}{3}$.

Solve $20\frac{5}{7} - 4\frac{1}{7}$.

2. 20 - 4 = 16

3. 5 - 1 = 4

$20\frac{5}{7} - 4\frac{1}{7} = 16\frac{4}{7}$

$16\frac{4}{7}$

1. Borrow If Necessary.

The Numerator of the Subtrahend is not larger than the Numerator of the Minuend, therefore it is not necessary to borrow.

2. Subtract the second Whole Number from the first.

20 - 4 = 16

3. Subtract the second Numerator from the first Numerator. The result will become the answer's Numerator. The first Numerator is 5 and the second Numerator is 1.

5 - 1 = 4

The answer's Denominator will be the same as the Denominator in the fractions that were in the subtraction problem.

The Denominator in both fractions that were in the subtraction problem is 7, so 7 will be the Denominator in the answer.

4. Reduce the answer's fraction part to its Lowest Terms if possible.

The fraction part is $\frac{4}{7}$.

$\frac{4}{7}$ is already in its Lowest Terms.

$16\frac{4}{7}$ is the answer to this problem.

Solve $11\frac{13}{16} - 6\frac{5}{16}$.

2. 11 - 6 = 5

3. 13 - 5 = 8

$11\frac{13}{16} - 6\frac{5}{16} = 5\frac{8}{16}$

4. $5\frac{8}{16} = 5\frac{1}{2}$

$5\frac{1}{2}$

1. Borrow If Necessary.

The Numerator of the Subtrahend is not larger than the Numerator of the Minuend, therefore it is not necessary to borrow.

2. Subtract the second Whole Number from the first.

11 - 6 = 5

3. Subtract the second Numerator from the first Numerator. The result will become the answer's Numerator. The first Numerator is 13 and the second Numerator is 5.

13 - 5 = 8

The answer's Denominator will be the same as the Denominator in the fractions that were in the subtraction problem.

The Denominator in both fractions that were in the subtraction problem is 16, so 16 will be the Denominator in the answer.

4. Reduce the answer's fraction part to its Lowest Terms if possible.

The fraction part is $\frac{8}{16}$.

$\frac{8}{16}$ reduced to its Lowest Terms is $\frac{1}{2}$.

$5\frac{1}{2}$ is the answer to this problem.

The procedures for subtracting Mixed Numbers with different Denominators are below:

Step 1. Change the fractions so that they have the same Denominator by finding the Lowest Common Denominator of all fractions involved.

Step 2. Borrow if necessary.

Step 3. Subtract the second Whole Number from the first Whole Number.

Step 4. Subtract the second Numerator from the first Numerator.

Step 5. If there is a fraction in the answer, reduce it to its Lowest Terms if possible.

Note: You can subtract the Numerators before subtracting the Whole Numbers if you want.

Example: $17\frac{9}{10} - 10\frac{3}{6}$

Step 1. The Lowest Common Denominator of both fractions is 30. This changes both fractions.

$\frac{9}{10} = \frac{27}{30}$ and $\frac{3}{6} = \frac{15}{30}$

Step 2. The Numerator of the Subtrahend is not larger than the Numerator of the Minuend, therefore it is not necessary to borrow.

Step 3. Subtract the second Whole Number from the first Whole Number.

17 – 10 = 7

$$\begin{array}{rcl} 17\frac{9}{10} & = & 17\frac{27}{30} \\ -\ 10\frac{3}{6} & = & 10\frac{15}{30} \\ \hline & = & 7 \end{array}$$

Step 4. Subtract the second Numerator from the first Numerator.

27 – 15 = 12

$$\begin{array}{rcl} 17\frac{9}{10} & = & 17\frac{27}{30} \\ -\ 10\frac{3}{6} & = & 10\frac{15}{30} \\ \hline & = & 7\frac{12}{30} \end{array}$$

Step 5. Reduce the answer's fraction to its Lowest Terms if possible.

The fraction part of $7\frac{12}{30}$ is $\frac{12}{30}$.

The GCF of the Numerator 12 and the Denominator 30 is 6.

$12 \div 6 = 2$
$30 \div 6 = 5$

The Lowest Terms of $\frac{12}{30}$ is $\frac{2}{5}$.

The answer to this problem is $7\frac{2}{5}$.

Solve $9\frac{2}{3} - 6\frac{11}{18}$.

1. $9\frac{2}{3} = 9\frac{12}{18}$ and $6\frac{11}{18} = 6\frac{11}{18}$

$9\frac{2}{3} - 6\frac{11}{18} = 9\frac{12}{18} - 6\frac{11}{18}$

3. 9 - 6 = 3

4. 12 - 11 = 1

$$
\begin{array}{r}
9\frac{2}{3} = 9\frac{12}{18} \\
-\ 6\frac{11}{18} = 6\frac{11}{18} \\
\hline
= 3\frac{1}{18}
\end{array}
$$

$\mathbf{3\frac{1}{18}}$

1. Change the fractions so that they have the same Denominator by finding the Lowest Common Denominator of all fractions involved.

$\frac{2}{3}$ changes to $\frac{12}{18}$ and $\frac{11}{18}$ stays at $\frac{11}{18}$.

2. Borrow If Necessary.

The Numerator of the Subtrahend is not larger than the Numerator of the Minuend, therefore it is not necessary to borrow.

3. Subtract the second Whole Number from the first.

9 - 6 = 3

4. Subtract the second Numerator from the first Numerator. The result will become the answer's Numerator. The first Numerator is 12 and the second Numerator is 11.

12 - 11 = 1

The answer's Denominator will be the same as the Denominator in the fractions that were used for the actual subtraction.

The Denominator in both fractions that were used for the actual subtraction is 18, so 18 will be the Denominator in the answer.

5. Reduce the answer's fraction part to its Lowest Terms if possible.

The fraction part is $\frac{1}{18}$.

$\frac{1}{18}$ is already in its Lowest Terms.

$3\frac{1}{18}$ is the answer to this problem.

Solve $18\frac{10}{13} - 10\frac{7}{26}$.

1. $18\frac{10}{13} = 18\frac{20}{26}$ and $10\frac{7}{26} = 10\frac{7}{26}$

$$18\frac{10}{13} - 10\frac{7}{26} = 18\frac{20}{26} - 10\frac{7}{26}$$

3. $18 - 10 = 8$

4. $20 - 7 = 13$

$$\begin{array}{rl} & 18\frac{10}{13} = 18\frac{20}{26} \\ - & 10\frac{7}{26} = 10\frac{7}{26} \\ \hline & \quad = 8\frac{13}{26} \end{array}$$

5. $8\frac{13}{26} = 8\frac{1}{2}$

$8\frac{1}{2}$

1. Change the fractions so that they have the same Denominator by finding the Lowest Common Denominator of all fractions involved.

   $\frac{10}{13}$ changes to $\frac{20}{26}$ and $\frac{7}{26}$ stays at $\frac{7}{26}$.

2. Borrow If Necessary.

   The Numerator of the Subtrahend is not larger than the Numerator of the Minuend, therefore it is not necessary to borrow.

3. Subtract the second Whole Number from the first.

   18 - 10 = 8

4. Subtract the second Numerator from the first Numerator. The result will become the answer's Numerator. The first Numerator is 20 and the second Numerator is 7.

   20 - 7 = 13

   The answer's Denominator will be the same as the Denominator in the fractions that were used for the actual subtraction.

   The Denominator in both fractions that were used for the actual subtraction is 26, so 26 will be the Denominator in the answer.

5. Reduce the answer's fraction part to its Lowest Terms if possible.

   The fraction part is $\frac{13}{26}$.

   $\frac{13}{26}$ reduced to its Lowest Terms is $\frac{1}{2}$.

$8\frac{1}{2}$ **is the answer to this problem.**

Solve $8\frac{1}{2} - 5\frac{3}{4}$.

1. $8\frac{1}{2} = 8\frac{2}{4}$ and $5\frac{3}{4} = 5\frac{3}{4}$

$8\frac{1}{2} - 5\frac{3}{4} = 8\frac{2}{4} - 5\frac{3}{4}$

2. $8\frac{1}{2} = \overset{7}{\cancel{8}}\frac{\overset{6}{\cancel{2}}}{4}$

$- \; 5\frac{3}{4} = 5\frac{3}{4}$

3. 7 - 5 = 2

4. 6 - 3 = 3

$8\frac{1}{2} = \overset{7}{\cancel{8}}\frac{\overset{6}{\cancel{2}}}{4}$

$- \; 5\frac{3}{4} = 5\frac{3}{4}$

$= 2\frac{3}{4}$

$2\frac{3}{4}$

1. Change the fractions so that they have the same Denominator by finding the Lowest Common Denominator of all fractions involved.

$\frac{1}{2}$ changes to $\frac{2}{4}$ and $\frac{3}{4}$ stays at $\frac{3}{4}$.

2. Borrow If Necessary.

The Numerator of the Subtrahend is larger than the Numerator of the Minuend, therefore we will borrow 1 from the Whole Number part of the Minuend.

8 - 1 = 7

Next, we will add the 1 that was borrowed to the fraction of the Minuend in the form of Denominator over the Denominator. The Numerator will be changed by this action.

$\frac{2}{4} + \frac{4}{4} = \frac{6}{4}$

3. Subtract the second Whole Number from the first.

7 - 5 = 2

4. Subtract the second Numerator from the first Numerator. The result will become the answer's Numerator. The first Numerator is 6 and the second Numerator is 3.

6 - 3 = 3

The answer's Denominator will be the same as the Denominator in the fractions that were used for the actual subtraction.

The Denominator in both fractions that were used for the actual subtraction is 4, so 4 will be the Denominator in the answer.

5. Reduce the answer's fraction part to its Lowest Terms if possible.

The fraction part is $\frac{3}{4}$.

$\frac{3}{4}$ is already in its Lowest Terms.

$2\frac{3}{4}$ is the answer to this problem.

Solve $15\frac{2}{4} - 3\frac{5}{6}$.

1. $15\frac{2}{4} = 15\frac{6}{12}$ and $3\frac{5}{6} = 3\frac{10}{12}$

$15\frac{2}{4} - 3\frac{5}{6} = 15\frac{6}{12} - 3\frac{10}{12}$

2. $15\frac{2}{4} = \overset{14}{\cancel{15}}\frac{\overset{18}{\cancel{6}}}{12}$

$-\ 3\frac{5}{6} = 3\frac{10}{12}$

3. 14 - 3 = 11

4. 18 - 10 = 8

$15\frac{2}{4} = \overset{14}{\cancel{15}}\frac{\overset{18}{\cancel{6}}}{12}$

$-\ 3\frac{5}{6} = 3\frac{10}{12}$

$= 11\frac{8}{12}$

5. $11\frac{8}{12} = 11\frac{2}{3}$

$11\frac{2}{3}$

---

1. Change the fractions so that they have the same Denominator by finding the Lowest Common Denominator of all fractions involved.

$\frac{2}{4}$ changes to $\frac{6}{12}$, and $\frac{5}{6}$ changes to $\frac{10}{12}$.

2. Borrow If Necessary.

The Numerator of the Subtrahend is larger than the Numerator of the Minuend, therefore we will borrow 1 from the Whole Number part of the Minuend.

15 - 1 = 14

Next, we will add the 1 that was borrowed to the fraction of the Minuend in the form of Denominator over the Denominator. The Numerator will be changed by this action.

$\frac{6}{12} + \frac{12}{12} = \frac{18}{12}$

3. Subtract the second Whole Number from the first.

14 - 3 = 11

4. Subtract the second Numerator from the first Numerator. The result will become the answer's Numerator. The first Numerator is 18 and the second Numerator is 10.

18 - 10 = 8

The answer's Denominator will be the same as the Denominator in the fractions that were used for the actual subtraction.

The Denominator in both fractions that were used for the actual subtraction is 12, so 12 will be the Denominator in the answer.

5. Reduce the answer's fraction part to its Lowest Terms if possible.

The fraction part is $\frac{8}{12}$.

$\frac{8}{12}$ reduced to its Lowest Terms is $\frac{2}{3}$.

$11\frac{2}{3}$ is the answer to this problem.

When multiplying fractions you can use Cancellation to reduce the fractions involved so that the result of the multiplication is not so large. Using Cancellation reduces the size of the answer, thus saving you work in reducing the answer to its Lowest Terms.

Cancellation involves finding a Greatest Common Factor that will divide into the Numerator of one fraction and the Denominator of the other fraction so that a reduction occurs in the Numerator and Denominator.

The procedures for using Cancellation are below:

Step 1. See if the Numerator of the first fraction and the Denominator of the second fraction have a Greatest Common Factor other than the number 1.

- ***A.*** If there is not a Greatest Common Factor other than 1, then the Numerator of the first fraction and the Denominator of the second fraction cannot be reduced.
- ***B.*** If there is a Greatest Common Factor other than 1.
  - ***B1.*** Divide the Numerator by the Greatest Common Factor. The result replaces the Numerator in the first fraction.
  - ***B2.*** Divide the Denominator by the Greatest Common Factor. The result replaces the Denominator in the second fraction.

Step 2. See if the Numerator of the second fraction and the Denominator of the first fraction have a Greatest Common Factor other than the number 1.

- ***A.*** If there is not a Greatest Common Factor other than 1, then the Numerator of the second fraction and the Denominator of the first fraction cannot be reduced.
- ***B.*** If there is a Greatest Common Factor other than 1.
  - ***B1.*** Divide the Numerator by the Greatest Common Factor. The result replaces the Numerator in the second fraction.
  - ***B2.*** Divide the Denominator by the Greatest Common Factor. The result replaces the Denominator in the first fraction.

The Math Mechanic Series: Fractions & Mixed Numbers Edition
Cancellation for Multiplying Fractions - Pg. 72

Example: Use Cancellation on $\frac{11}{96} \times \frac{8}{44}$.

Step 1. See if the Numerator of the first fraction and the Denominator of the second fraction have a GCF other than 1.
The Numerator of the first fraction is 11 and the Denominator of the second fraction is 44.

B. The GCF of 11 and 44 is 11.
B1. Divide the Numerator by the Greatest Common Factor. The result replaces the Numerator.

$11 \div 11 = 1$ 1 Replaces 11 in the Numerator.

$$\frac{\overset{1}{\cancel{11}}}{96} \times \frac{8}{44}$$

B2. Divide the Denominator by the Greatest Common Factor. The result replaces the Denominator.

$44 \div 11 = 4$ 4 Replaces 44 in the Denominator.

$$\frac{\overset{1}{\cancel{11}}}{96} \times \frac{8}{\underset{4}{\cancel{44}}}$$

Step 2. See if the Numerator of the second fraction and the Denominator of the first fraction have a GCF other than 1.
The Numerator of the second fraction is 8 and the Denominator of the first fraction is 96.

B. The GCF of 8 and 96 is 8.
B1. Divide the Numerator by the Greatest Common Factor. The result replaces the Numerator.

$8 \div 8 = 1$ 1 Replaces 8 in the Numerator.

$$\frac{\overset{1}{\cancel{11}}}{96} \times \frac{\overset{1}{\cancel{8}}}{\underset{4}{\cancel{44}}}$$

B2. Divide the Denominator by the Greatest Common Factor. The result replaces the Denominator.

$96 \div 8 = 12$ 12 Replaces 96 in the Denominator.

$$\frac{\overset{1}{\cancel{11}}}{\underset{12}{\cancel{96}}} \times \frac{\overset{1}{\cancel{8}}}{\underset{4}{\cancel{44}}}$$

Cancellation changes $\frac{11}{96} \times \frac{8}{44}$ to $\frac{1}{12} \times \frac{1}{4}$.

The procedures for multiplying fractions are below:

Step 1. Reduce the fractions using Cancellation if possible.
Step 2. Multiply the Numerators.
Step 3. Multiply the Denominators.
Step 4. If the answer is an Improper Fraction, change it to a Mixed Number or a Whole Number.
Step 5. If there is a fraction in the answer, reduce it to its Lowest Terms if possible.

Example: $\frac{3}{8} \times \frac{1}{5}$

Step 1. Reduce the fractions using Cancellation if possible.

The Numerator of the first fraction which is 3 and the Denominator of the second fraction which is 5 do not have a common factor other than 1, so they cannot be cancelled.

The Numerator of the second fraction which is 1 and the Denominator of the first fraction which is 8 do not have a common factor other than 1, so they cannot be cancelled.

The problem will remain $\frac{3}{8} \times \frac{1}{5}$.

Step 2. Multiply the Numerators.

$\frac{3}{8} \times \frac{1}{5}$

3 × 1 = 3

Step 3. Multiply the Denominators.

$\frac{3}{8} \times \frac{1}{5}$

8 × 5 = 40

$\frac{3}{8} \times \frac{1}{5} = \frac{3}{40}$

Step 4. If the answer is an Improper Fraction, change it to a Mixed Number or a Whole Number.

$\frac{3}{40}$ is not an Improper Fraction.

Step 5. If there is a fraction in the answer, reduce it to its Lowest Terms if possible.

$\frac{3}{40}$ is already in its Lowest Terms.

The answer to this problem is $\frac{3}{40}$.

Solve $\frac{1}{5} \times \frac{2}{3}$.

2. $1 \times 2 = 2$

3. $5 \times 3 = 15$

$\frac{1}{5} \times \frac{2}{3} = \frac{2}{15}$

$\mathbf{\frac{2}{15}}$

1. Reduce the fractions using Cancellation if possible.

This problem cannot be reduced by Cancellation.

2. Multiply the Numerators.

The Numerators are 1 and 2. The result will become the answer's Numerator.

$1 \times 2 = 2$

3. Multiply the Denominators.

The Denominators are 5 and 3. The result will become the answer's Denominator.

$5 \times 3 = 15$

4. If the answer is an Improper Fraction, change it to a Mixed Number or a Whole Number.

$\frac{2}{15}$ **is not an Improper Fraction.**

5. If there is a fraction in the answer, reduce it to its Lowest Terms if possible.

$\frac{2}{15}$ **is already in its Lowest Terms.**

$\frac{2}{15}$ **is the answer to this problem.**

The Math Mechanic Series: Fractions & Mixed Numbers Edition
Step-by-Step Examples of Multiplying Fractions - Pg. 75

Solve $\frac{4}{5} \times \frac{3}{9}$.

2. $4 \times 3 = 12$

3. $5 \times 9 = 45$

5. $\frac{12}{45} = \frac{4}{15}$

$\frac{4}{15}$

---

1. Reduce the fractions using Cancellation if possible.

This problem cannot be reduced by Cancellation.

2. Multiply the Numerators.

The Numerators are 4 and 3. The result will become the answer's Numerator.

$4 \times 3 = 12$

3. Multiply the Denominators.

The Denominators are 5 and 9. The result will become the answer's Denominator.

$5 \times 9 = 45$

4. If the answer is an Improper Fraction, change it to a Mixed Number or a Whole Number.

$\frac{12}{45}$ **is not an Improper Fraction.**

5. If there is a fraction in the answer, reduce it to its Lowest Terms if possible.

$\frac{12}{45}$ **reduced to its Lowest Terms is** $\frac{4}{15}$.

$\frac{4}{15}$ **is the answer to this problem.**

Solve $\frac{6}{8} \times \frac{4}{10}$.

1. 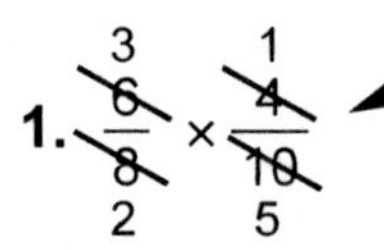

$\frac{3}{2} \times \frac{1}{5}$

2. $3 \times 1 = 3$

3. $2 \times 5 = 10$

$\frac{3}{2} \times \frac{1}{5} = \frac{3}{10}$

$\frac{3}{10}$

1. Reduce the fractions using Cancellation if possible.

$\frac{6}{8} \times \frac{4}{10}$ **reduces to** $\frac{3}{2} \times \frac{1}{5}$.

2. Multiply the Numerators.

The Numerators are 3 and 1. The result will become the answer's Numerator.

$3 \times 1 = 3$

3. Multiply the Denominators.

The Denominators are 2 and 5. The result will become the answer's Denominator.

$2 \times 5 = 10$

4. If the answer is an Improper Fraction, change it to a Mixed Number or a Whole Number.

$\frac{3}{10}$ **is not an Improper Fraction.**

5. If there is a fraction in the answer, reduce it to its Lowest Terms if possible.

$\frac{3}{10}$ **is already in its Lowest Terms.**

$\frac{3}{10}$ **is the answer to this problem.**

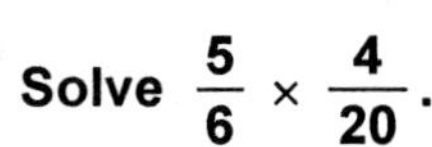

Solve $\frac{5}{6} \times \frac{4}{20}$.

1. $\frac{\overset{1}{\cancel{5}}}{\underset{3}{\cancel{6}}} \times \frac{\overset{2}{\cancel{4}}}{\underset{4}{\cancel{20}}}$

$\frac{1}{3} \times \frac{2}{4}$

2. $1 \times 2 = 2$

3. $3 \times 4 = 12$

$\frac{1}{3} \times \frac{2}{4} = \frac{2}{12}$

5. $\frac{2}{12} = \frac{1}{6}$

$\frac{1}{6}$

1. Reduce the fractions using Cancellation if possible.

$\frac{5}{6} \times \frac{4}{20}$ reduces to $\frac{1}{3} \times \frac{2}{4}$.

2. Multiply the Numerators.

The Numerators are 1 and 2. The result will become the answer's Numerator.

$1 \times 2 = 2$

3. Multiply the Denominators.

The Denominators are 3 and 4. The result will become the answer's Denominator.

$3 \times 4 = 12$

4. If the answer is an Improper Fraction, change it to a Mixed Number or a Whole Number.

$\frac{2}{12}$ is not an Improper Fraction.

5. If there is a fraction in the answer, reduce it to its Lowest Terms if possible.

$\frac{2}{12}$ reduced to its Lowest Terms is $\frac{1}{6}$.

$\frac{1}{6}$ is the answer to this problem.

The procedures for multiplying Mixed Numbers are below:
Step 1. Convert all Mixed Numbers to Improper Fractions.
Step 2. Reduce the fractions using Cancellation if possible.
Step 3. Multiply the Numerators.
Step 4. Multiply the Denominators.
Step 5. If the answer is an Improper Fraction, change it to a Mixed Number or a Whole Number.
Step 6. If there is a fraction in the answer, reduce it to its Lowest Terms if possible.

Example: $3\frac{2}{3} \times 1\frac{2}{3}$

Step 1. Convert all Mixed Numbers to Improper Fractions.

$3\frac{2}{3}$ converts to $\frac{11}{3}$

$1\frac{2}{3}$ converts to $\frac{5}{3}$

The conversion changes the problem to $\frac{11}{3} \times \frac{5}{3}$.

Step 2. Reduce the fractions using Cancellation if possible.

The Numerator of $\frac{11}{3}$ which is 11 and the Denominator of $\frac{5}{3}$ which 3 is do not have a greatest common factor other than 1, therefore they cannot be reduced.

The Denominator of $\frac{11}{3}$ which is 3 and the Numerator of $\frac{5}{3}$ which is 5, do not have a greatest common factor other than 1, therefore they cannot be reduced.

The problem remains $\frac{11}{3} \times \frac{5}{3}$

Step 3. Multiply the Numerators.

$\frac{11}{3} \times \frac{5}{3}$

11 × 5 = 55

Step 4. Multiply the Denominators.

$\frac{11}{3} \times \frac{5}{3}$

3 × 3 = 9

$\frac{11}{3} \times \frac{5}{3} = \frac{55}{9}$

Step 5. If the answer is an Improper Fraction, change it to a Mixed Number or a Whole Number.

$\frac{55}{9}$ is an improper fraction so it should be converted to a Mixed Number.

$\frac{55}{9}$ converts to $6\frac{1}{9}$.

Step 6. If there is a fraction in the answer, reduce it to its Lowest Terms if possible.

The fraction part of $6\frac{1}{9}$ is $\frac{1}{9}$. This fraction is already in its lowest terms.

The answer to this problem is $6\frac{1}{9}$.

Solve $4\frac{6}{7} \times \frac{1}{2}$.

1. $4\frac{6}{7} = \frac{34}{7}$

$\frac{34}{7} \times \frac{1}{2}$

2. $\frac{\cancel{34}^{17}}{7} \times \frac{1}{\cancel{2}_{1}}$

$\frac{17}{7} \times \frac{1}{1}$

3. $17 \times 1 = 17$

4. $7 \times 1 = 7$

$\frac{17}{7} \times \frac{1}{1} = \frac{17}{7}$

5. $\frac{17}{7} = 2\frac{3}{7}$

$2\frac{3}{7}$

1. Convert all Mixed Numbers to Improper Fractions.

$4\frac{6}{7}$ is a Mixed Number so it will be changed to an Improper Fraction.

$4\frac{6}{7}$ changes to $\frac{34}{7}$.

2. Reduce the fractions using Cancellation if possible.

$\frac{34}{7} \times \frac{1}{2}$ reduces to $\frac{17}{7} \times \frac{1}{1}$.

3. Multiply the Numerators.

The Numerators are 17 and 1. The result will become the answer's Numerator.

$17 \times 1 = 17$

4. Multiply the Denominators.

The Denominators are 7 and 1. The result will become the answer's Denominator.

$7 \times 1 = 7$

5. If the answer is an Improper Fraction, change it to a Mixed Number or a Whole Number.

$\frac{17}{7}$ is an Improper Fraction, so it will be changed to a Mixed Number.

$\frac{17}{7} = 2\frac{3}{7}$

6. If there is a fraction in the answer, reduce it to its Lowest Terms if possible.

$\frac{3}{7}$ is already in its Lowest Terms.

$2\frac{3}{7}$ is the answer to this problem.

Solve $1\frac{5}{9} \times 2\frac{1}{4}$.

1. $1\frac{5}{9} = \frac{14}{9}$ and $2\frac{1}{4} = \frac{9}{4}$

$\frac{14}{9} \times \frac{9}{4}$

2. $\frac{\cancel{14}^{7}}{\cancel{9}_{1}} \times \frac{\cancel{9}^{1}}{\cancel{4}_{2}}$

$\frac{7}{1} \times \frac{1}{2}$

3. $7 \times 1 = 7$

4. $1 \times 2 = 2$

$\frac{7}{1} \times \frac{1}{2} = \frac{7}{2}$

5. $\frac{7}{2} = 3\frac{1}{2}$

$\mathbf{3\frac{1}{2}}$

1. Convert all Mixed Numbers to Improper Fractions.

$1\frac{5}{9}$ and $2\frac{1}{4}$ are Mixed Numbers so they will be changed to Improper Fractions.

$1\frac{5}{9}$ changes to $\frac{14}{9}$ and $2\frac{1}{4}$ changes to $\frac{9}{4}$.

2. Reduce the fractions using Cancellation if possible.

$\frac{14}{9} \times \frac{9}{4}$ reduces to $\frac{7}{1} \times \frac{1}{2}$.

3. Multiply the Numerators.

The Numerators are 7 and 1. The result will become the answer's Numerator.

$7 \times 1 = 7$

4. Multiply the Denominators.

The Denominators are 1 and 2. The result will become the answer's Denominator.

$1 \times 2 = 2$

5. If the answer is an Improper Fraction, change it to a Mixed Number or a Whole Number.

$\frac{7}{2}$ is an Improper Fraction, so it will be changed to a Mixed Number.

$\frac{7}{2} = 3\frac{1}{2}$

6. If there is a fraction in the answer, reduce it to its Lowest Terms if possible.

$\frac{1}{2}$ is already in its Lowest Terms.

$3\frac{1}{2}$ is the answer to this problem.

The procedure for reciprocating a fraction is as follows:

Switch the Numerator and the Denominator.

Example: The reciprocal of $\frac{9}{10}$ is $\frac{10}{9}$.

Example: The reciprocal of $\frac{3}{4}$ is $\frac{4}{3}$.

Example: The reciprocal of $\frac{10}{7}$ is $\frac{7}{10}$.

Example: The reciprocal of $\frac{8}{16}$ is $\frac{16}{8}$.

Example: The reciprocal of $\frac{5}{2}$ is $\frac{2}{5}$.

The procedures for dividing one fraction by another are below:

Step 1. Reciprocate the fraction that comes after the '÷' sign and change the '÷' sign to a '×' sign.
Step 2. Reduce the fractions using Cancellation if possible.
Step 3. Multiply the Numerators.
Step 4. Multiply the Denominators.
Step 5. If the answer is an Improper Fraction, change it to a Mixed Number or a Whole Number.
Step 6. If there is a fraction in the answer, reduce it to its Lowest Terms if possible.

See the next 2 pages for Step-by-Step Examples.

Solve $\frac{3}{4} \div \frac{2}{5}$.

1. $\frac{3}{4} \div \frac{2}{5} = \frac{3}{4} \times \frac{5}{2}$

3. $3 \times 5 = 15$

4. $4 \times 2 = 8$

$\frac{3}{4} \times \frac{5}{2} = \frac{15}{8}$

5. $\frac{15}{8} = 1\frac{7}{8}$

$1\frac{7}{8}$

1. Reciprocate the fraction that comes after the '÷' sign and change the '÷' sign to a 'x' sign.

$\frac{2}{5}$ becomes $\frac{5}{2}$ and the '÷' sign changes to the 'x' sign.

2. Reduce the fractions using Cancellation if possible.

This problem cannot be reduced by Cancellation.

3. Multiply the Numerators.

The Numerators are 3 and 5. The result will become the answer's Numerator.

$3 \times 5 = 15$

4. Multiply the Denominators.

The Denominators are 4 and 2. The result will become the answer's Denominator.

$4 \times 2 = 8$

5. If the answer is an Improper Fraction, change it to a Mixed Number or a Whole Number.

$\frac{15}{8}$ is an Improper Fraction, so it will be changed to a Mixed Number.

$\frac{15}{8} = 1\frac{7}{8}$

6. If there is a fraction in the answer, reduce it to its Lowest Terms if possible.

$\frac{7}{8}$ is already in its Lowest Terms.

$1\frac{7}{8}$ is the answer to this problem.

Solve $\mathbf{\frac{5}{25} \div \frac{10}{30}}$.

1. $\frac{5}{25} \div \frac{10}{30} = \frac{5}{25} \times \frac{30}{10}$

2. $\frac{\cancel{5}^{1}}{\cancel{25}_{5}} \times \frac{\cancel{30}^{6}}{\cancel{10}_{2}}$

$\frac{1}{5} \times \frac{6}{2}$

3. $1 \times 6 = 6$

4. $5 \times 2 = 10$

$\frac{1}{5} \times \frac{6}{2} = \frac{6}{10}$

6. $\frac{6}{10} = \frac{3}{5}$

$\mathbf{\frac{3}{5}}$

1. Reciprocate the fraction that comes after the '÷' sign and change the '÷' sign to a '×' sign.

$\frac{10}{30}$ **becomes** $\frac{30}{10}$ **and the '÷' sign changes to the '×' sign.**

2. Reduce the fractions using Cancellation if possible.

$\frac{5}{25} \times \frac{30}{10}$ **reduces to** $\frac{1}{5} \times \frac{6}{2}$.

3. Multiply the Numerators.

The Numerators are 1 and 6. The result will become the answer's Numerator.

$\mathbf{1 \times 6 = 6}$

4. Multiply the Denominators.

The Denominators are 5 and 2. The result will become the answer's Denominator.

$\mathbf{5 \times 2 = 10}$

5. If the answer is an Improper Fraction, change it to a Mixed Number or a Whole Number.

$\frac{6}{10}$ **is not an Improper Fraction.**

6. If there is a fraction in the answer, reduce it to its Lowest Terms if possible.

$\frac{6}{10}$ **reduced to its Lowest Terms is** $\frac{3}{5}$.

$\frac{3}{5}$ **is the answer to this problem.**

The procedures for division involving Mixed Numbers are below:

Step 1. Convert all Mixed Numbers to Improper Fractions.
Step 2. Reciprocate the fraction that comes after the '÷' sign and change the '÷' sign to a '×' sign.
Step 3. Reduce the fractions using Cancellation if possible.
Step 4. Multiply the Numerators.
Step 5. Multiply the Denominators.
Step 6. If the answer is an Improper Fraction, change it to a Mixed Number or a Whole Number.
Step 7. If there is a fraction in the answer, reduce it to its Lowest Terms if possible.

See the next 2 pages for Step-by-Step Examples.

Solve $5\frac{1}{2} \div 3\frac{1}{3}$.

1. $5\frac{1}{2} = \frac{11}{2}$ and $3\frac{1}{3} = \frac{10}{3}$

The problem now becomes $\frac{11}{2} \div \frac{10}{3}$

2. $\frac{11}{2} \div \frac{10}{3} = \frac{11}{2} \times \frac{3}{10}$

4. $11 \times 3 = 33$

5. $2 \times 10 = 20$

$\frac{11}{2} \times \frac{3}{10} = \frac{33}{20}$

6. $\frac{33}{20} = 1\frac{13}{20}$

$1\frac{13}{20}$

1. Convert all Mixed Numbers to Improper Fractions.

$5\frac{1}{2}$ and $3\frac{1}{3}$ are Mixed Numbers so they will be changed to Improper Fractions.

$5\frac{1}{2}$ changes to $\frac{11}{2}$ and $3\frac{1}{3}$ changes to $\frac{10}{3}$.

2. Reciprocate the fraction that comes after the '÷' sign and change the '÷' sign to a '×' sign.

$\frac{10}{3}$ becomes $\frac{3}{10}$ and the '÷' sign changes to the '×' sign.

3. Reduce the fractions using Cancellation if possible.

This problem cannot be reduced by Cancellation.

4. Multiply the Numerators.

The Numerators are 11 and 3. The result will become the answer's Numerator.

$11 \times 3 = 33$

5. Multiply the Denominators.

The Denominators are 2 and 10. The result will become the answer's Denominator.

$2 \times 10 = 20$

6. If the answer is an Improper Fraction, change it to a Mixed Number or a Whole Number.

$\frac{33}{20}$ is an Improper Fraction, so it will be changed to a Mixed Number.

$\frac{33}{20} = 1\frac{13}{20}$

7. If there is a fraction in the answer, reduce it to its Lowest Terms if possible.

$\frac{13}{20}$ is already in its Lowest Terms.

$1\frac{13}{20}$ is the answer to this problem.

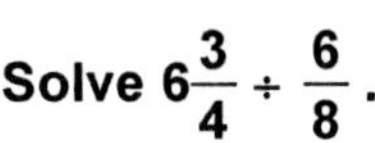

Solve $6\frac{3}{4} \div \frac{6}{8}$.

1. $6\frac{3}{4} = \frac{27}{4}$

$\frac{27}{4} \div \frac{6}{8}$

2. $\frac{27}{4} \div \frac{6}{8} = \frac{27}{4} \times \frac{8}{6}$

3. $\frac{\cancel{27}^{9}}{\cancel{4}_{1}} \times \frac{\cancel{8}^{2}}{\cancel{6}_{2}}$

$\frac{9}{1} \times \frac{2}{2}$

4. $9 \times 2 = 18$

5. $1 \times 2 = 2$

$\frac{9}{1} \times \frac{2}{2} = \frac{18}{2}$

6. $\frac{18}{2} = 9$

9

1. Convert all Mixed Numbers to Improper Fractions.

$6\frac{3}{4}$ is a Mixed Numbers so it will be changed to an Improper Fraction.

$6\frac{3}{4}$ changes to $\frac{27}{4}$.

2. Reciprocate the fraction that comes after the '÷' sign and change the '÷' sign to a '×' sign.

$\frac{6}{8}$ becomes $\frac{8}{6}$ and the '÷' sign changes to the '×' sign.

3. Reduce the fractions using Cancellation if possible.

$\frac{27}{4} \times \frac{8}{6}$ reduces to $\frac{9}{1} \times \frac{2}{2}$.

4. Multiply the Numerators.

The Numerators are 9 and 2. The result will become the answer's Numerator.

$9 \times 2 = 18$

5. Multiply the Denominators.

The Denominators are 1 and 2. The result will become the answer's Denominator.

$1 \times 2 = 2$

6. If the answer is an Improper Fraction, change it to a Mixed Number or a Whole Number.

$\frac{18}{2}$ is an Improper Fraction, so it will be changed to a Whole Number since 18 ÷ 2 does not have a remainder.

$\frac{18}{2} = 9$

7. If there is a fraction in the answer, reduce it to its Lowest Terms if possible.

There is no fraction in the answer. 9 is a Whole Number, so it cannot be reduced.

9 is the answer to this problem.

Helpful Hints

1. If the Numerator and Denominator of an answer are both even numbers, the fraction can be reduced to Lower Terms. The reason is because both numbers will at least have 2 as a Common Factor.

2. If an answer contains a fraction and that fraction's Numerator is 0, the fraction can be reduced to its Lowest Terms by dividing 0 by the Denominator. The result will always be 0, since 0 divided by any number is 0.

$\frac{0}{12}$ **reduces to 0 because** $0 \div 12 = 0$

3. If an answer is a mixed number and the fraction's Numerator part is 0, the fraction part can be reduced to its Lowest Terms by dividing 0 by the Denominator. The result will always be 0, since 0 divided by any number is 0. The answer would then change to that of the whole number, because you would add the 0 that resulted from reducing the fraction part to its Lowest Terms to the whole number.

$7\frac{0}{5}$ **changes to 7 after dividing 0 by 5 and adding the result which is 0 to 7**

4. If an answer contains a fraction and that fraction's Numerator and Denominator are the same number, the fraction can be reduced to its Lowest Terms by dividing the Numerator by the Denominator. The result will always be 1, since any number (with the exception of 0) that is divided by the same equivalent number equals 1.

$\frac{6}{6}$ **reduces to 1 because** $6 \div 6 = 1$

What If Section

What if there is an Improper Fraction at the beginning of a problem?

If there is an Improper Fraction in a problem that involves adding, subtracting, multiplying, or dividing, that problem is to be worked out in the same way that you would work out a problem that did not have an Improper Fraction.

Example: $\frac{7}{4} + \frac{5}{4}$

The answer to this problem is 3 and is worked in the same way as was stated on Page 43.

Section continued on next page.

What if there is a Whole Number at the beginning of a problem?

To add, subtract, multiply, or divide a Mixed Number or Fraction by a Whole Number, you should convert the Whole Number to an Improper Fraction, then proceed with the appropriate procedures that will solve the problem.

Example: $6 + \frac{1}{2}$

To solve this problem you would convert 6 to $\frac{6}{1}$.

The problem now becomes $\frac{6}{1} + \frac{1}{2}$.

Example: $14 - \frac{3}{4}$

To solve this problem you would convert 14 to $\frac{14}{1}$.

The problem now becomes $\frac{14}{1} - \frac{3}{4}$.

Example: $\frac{1}{3} \times 5$

To solve this problem you would convert 5 to $\frac{5}{1}$.

The problem now becomes $\frac{1}{3} \times \frac{5}{1}$.

Example: $8 \div \frac{1}{4}$

To solve this problem you would convert 8 to $\frac{8}{1}$.

The problem now becomes $\frac{8}{1} \div \frac{1}{4}$.

What if there is an Improper Fraction in the answer's fraction part in an adding or subtraction problem involving Mixed Numbers?

If this happens, covert the Improper Fraction to a Mixed Number or a Whole Number and add the Whole Number that resulted from the conversion to the Whole Number part of the answer.

Example: $2\frac{3}{2} + 5\frac{5}{2} = 7\frac{8}{2} = 11$

The answer above is $7\frac{8}{2}$ which is a Mixed Number with an Improper Fraction. $\frac{8}{2}$ converts to 4, so 4 will be added to 7 and the answer becomes 11.

Example: $4\frac{5}{3} + 10\frac{2}{3} = 14\frac{7}{3} = 16\frac{1}{3}$

The answer above is $14\frac{7}{3}$ which is a Mixed Number with an Improper Fraction. $\frac{7}{3}$ converts to $2\frac{1}{3}$, so 2 will be added to 14 and the answer becomes $16\frac{1}{3}$.

Example: $6\frac{7}{4} - 3\frac{1}{4} = 3\frac{6}{4} = 4\frac{2}{4} = 4\frac{1}{2}$

The answer above is $3\frac{6}{4}$ which is a Mixed Number with an Improper Fraction. $\frac{6}{4}$ converts to $1\frac{2}{4}$, so 1 will be added to 3 and the answer becomes $4\frac{2}{4}$. $4\frac{2}{4}$ can be reduced to $4\frac{1}{2}$.

Finding the Factors of a Number

Solve these problems on a separate sheet of paper since it is often difficult to work out problems in a book.

1. 1 **2.** 2 **3.** 4 **4.** 8

5. 3 **6.** 7 **7.** 10 **8.** 5

9. 6 **10.** 15 **11.** 16 **12.** 20

13. 50 **14.** 9 **15.** 18 **16.** 25

17. 12 **18.** 33 **19.** 27 **20.** 24

Finding the Greatest Common Factor

Solve these problems on a separate sheet of paper since it is often difficult to work out problems in a book.

1. 3 and 6
2. 2 and 4
3. 5 and 10
4. 14 and 28

5. 14 and 35
6. 8 and 20
7. 18 and 30
8. 20 and 15

9. 33 and 44
10. 6 and 2
11. 24 and 15
12. 16 and 24

13. 10 and 35
14. 14 and 18
15. 81 and 9
16. 15 and 21

17. 12 and 8
18. 55 and 30
19. 10 and 12
20. 6 and 12

Converting Whole Numbers to Improper Fractions

Solve these problems on a separate sheet of paper since it is often difficult to work out problems in a book.

1. 2 **2.** 6 **3.** 18 **4.** 8

5. 13 **6.** 10 **7.** 3 **8.** 25

9. 11 **10.** 16 **11.** 9 **12.** 58

13. 101 **14.** 67 **15.** 5 **16.** 20

17. 14 **18.** 7 **19.** 40 **20.** 4

The Math Mechanic Series: Fractions & Mixed Numbers Edition
Practice Problems - Pg. 93

Converting Improper Fractions to Whole Numbers

Solve these problems on a separate sheet of paper since it is often difficult to work out problems in a book.

1. $\frac{5}{1}$ **2.** $\frac{6}{3}$ **3.** $\frac{2}{1}$ **4.** $\frac{12}{4}$

5. $\frac{16}{8}$ **6.** $\frac{4}{1}$ **7.** $\frac{21}{7}$ **8.** $\frac{90}{10}$

9. $\frac{45}{5}$ **10.** $\frac{8}{2}$ **11.** $\frac{26}{13}$ **12.** $\frac{63}{9}$

13. $\frac{28}{14}$ **14.** $\frac{18}{3}$ **15.** $\frac{25}{5}$ **16.** $\frac{100}{50}$

17. $\frac{10}{2}$ **18.** $\frac{16}{4}$ **19.** $\frac{11}{1}$ **20.** $\frac{8}{1}$

The Math Mechanic Series: Fractions & Mixed Numbers Edition
Practice Problems - Pg. 94

Converting Improper Fractions to Mixed Numbers

Solve these problems on a separate sheet of paper since it is often difficult to work out problems in a book.

1. $\frac{8}{7}$ **2.** $\frac{3}{2}$ **3.** $\frac{9}{5}$ **4.** $\frac{13}{4}$

5. $\frac{18}{8}$ **6.** $\frac{19}{7}$ **7.** $\frac{20}{14}$ **8.** $\frac{15}{11}$

9. $\frac{7}{5}$ **10.** $\frac{35}{21}$ **11.** $\frac{25}{15}$ **12.** $\frac{33}{9}$

13. $\frac{6}{4}$ **14.** $\frac{17}{10}$ **15.** $\frac{9}{6}$ **16.** $\frac{16}{5}$

17. $\frac{22}{12}$ **18.** $\frac{30}{12}$ **19.** $\frac{41}{7}$ **20.** $\frac{6}{5}$

Converting Mixed Numbers to Improper Fractions

Solve these problems on a separate sheet of paper since it is often difficult to work out problems in a book.

1. $2\frac{1}{4}$ **2.** $1\frac{3}{4}$ **3.** $1\frac{8}{12}$ **4.** $5\frac{2}{3}$

5. $3\frac{3}{8}$ **6.** $1\frac{4}{7}$ **7.** $10\frac{1}{2}$ **8.** $4\frac{4}{5}$

9. $1\frac{3}{9}$ **10.** $6\frac{1}{3}$ **11.** $2\frac{1}{8}$ **12.** $7\frac{4}{6}$

13. $11\frac{1}{3}$ **14.** $4\frac{1}{2}$ **15.** $5\frac{7}{9}$ **16.** $3\frac{3}{4}$

17. $9\frac{2}{9}$ **18.** $3\frac{7}{8}$ **19.** $2\frac{3}{5}$ **20.** $8\frac{6}{11}$

Reducing a Fraction to its Lowest Terms

Solve these problems on a separate sheet of paper since it is often difficult to work out problems in a book.

1. $\frac{4}{8}$ **2.** $\frac{3}{9}$ **3.** $\frac{6}{18}$ **4.** $\frac{20}{25}$

5. $\frac{16}{22}$ **6.** $\frac{10}{18}$ **7.** $\frac{9}{24}$ **8.** $\frac{12}{33}$

9. $\frac{5}{25}$ **10.** $\frac{18}{36}$ **11.** $\frac{4}{6}$ **12.** $\frac{10}{45}$

13. $\frac{11}{66}$ **14.** $\frac{12}{28}$ **15.** $\frac{39}{63}$ **16.** $\frac{15}{60}$

17. $\frac{16}{40}$ **18.** $\frac{17}{51}$ **19.** $\frac{24}{60}$ **20.** $\frac{13}{52}$

The Math Mechanic Series: Fractions & Mixed Numbers Edition
Practice Problems - Pg. 97

Finding the Least Common Multiple

Solve these problems on a separate sheet of paper since it is often difficult to work out problems in a book.

1. 3 and 4	**2.** 1 and 8	**3.** 6 and 4	**4.** 2 and 5
5. 6 and 9	**6.** 3 and 7	**7.** 8 and 2	**8.** 2 and 1
9. 5 and 10	**10.** 12 and 3	**11.** 7 and 7	**12.** 4 and 5
13. 9 and 6	**14.** 2 and 3	**15.** 15 and 10	**16.** 1 and 1
17. 6 and 8	**18.** 9 and 3	**19.** 7 and 2	**20.** 5 and 3

Finding the Lowest Common Denominator

Solve these problems on a separate sheet of paper since it is often difficult to work out problems in a book.

1. $\frac{1}{4} + \frac{1}{2}$
2. $\frac{1}{3} + \frac{2}{6}$
3. $\frac{1}{2} + \frac{1}{2}$
4. $\frac{3}{4} + \frac{1}{8}$
5. $\frac{9}{15} - \frac{2}{5}$
6. $\frac{3}{8} - \frac{6}{24}$
7. $\frac{3}{10} - \frac{1}{5}$
8. $\frac{7}{10} - \frac{3}{10}$
9. $\frac{3}{4} + \frac{4}{5}$
10. $\frac{4}{9} + \frac{3}{6}$
11. $\frac{2}{3} - \frac{1}{8}$
12. $\frac{4}{7} - \frac{1}{4}$
13. $\frac{6}{9} - \frac{2}{18}$
14. $\frac{1}{3} + \frac{2}{9}$
15. $\frac{6}{27} + \frac{7}{9}$
16. $\frac{1}{4} - \frac{1}{14}$
17. $\frac{1}{3} + \frac{3}{4}$
18. $\frac{11}{19} - \frac{5}{38}$
19. $\frac{8}{10} + \frac{12}{15}$
20. $\frac{7}{8} - \frac{6}{7}$

Adding Fractions with the Same Denominator

Solve these problems on a separate sheet of paper since it is often difficult to work out problems in a book.

1. $\frac{1}{4}+\frac{2}{4}$ **2.** $\frac{3}{8}+\frac{4}{8}$ **3.** $\frac{1}{5}+\frac{1}{5}$ **4.** $\frac{1}{3}+\frac{1}{3}$

5. $\frac{3}{9}+\frac{4}{9}$ **6.** $\frac{4}{8}+\frac{2}{8}$ **7.** $\frac{1}{2}+\frac{1}{2}$ **8.** $\frac{3}{5}+\frac{3}{5}$

9. $\frac{1}{10}+\frac{4}{10}$ **10.** $\frac{5}{20}+\frac{8}{20}$ **11.** $\frac{9}{10}+\frac{6}{10}$ **12.** $\frac{2}{3}+\frac{1}{3}$

13. $\frac{4}{7}+\frac{4}{7}$ **14.** $\frac{1}{6}+\frac{3}{6}$ **15.** $\frac{4}{8}+\frac{6}{8}$ **16.** $\frac{3}{4}+\frac{2}{4}$

17. $\frac{1}{4}+\frac{1}{4}$ **18.** $\frac{5}{6}+\frac{4}{6}$ **19.** $\frac{3}{4}+\frac{1}{4}$ **20.** $\frac{5}{8}+\frac{1}{8}$

Adding Fractions with Different Denominators

Solve these problems on a separate sheet of paper since it is often difficult to work out problems in a book.

1. $\frac{3}{4} + \frac{1}{2}$

2. $\frac{2}{3} + \frac{3}{4}$

3. $\frac{9}{10} + \frac{2}{5}$

4. $\frac{2}{8} + \frac{1}{2}$

5. $\frac{3}{5} + \frac{1}{4}$

6. $\frac{8}{9} + \frac{5}{6}$

7. $\frac{4}{5} + \frac{4}{9}$

8. $\frac{3}{12} + \frac{1}{8}$

9. $\frac{2}{9} + \frac{5}{12}$

10. $\frac{1}{4} + \frac{1}{2}$

11. $\frac{6}{11} + \frac{1}{2}$

12. $\frac{3}{8} + \frac{15}{24}$

13. $\frac{7}{9} + \frac{5}{6}$

14. $\frac{3}{6} + \frac{4}{8}$

15. $\frac{9}{20} + \frac{3}{10}$

16. $\frac{4}{5} + \frac{1}{3}$

17. $\frac{13}{16} + \frac{7}{8}$

18. $\frac{7}{11} + \frac{2}{22}$

19. $\frac{6}{15} + \frac{8}{10}$

20. $\frac{1}{7} + \frac{1}{3}$

The Math Mechanic Series: Fractions & Mixed Numbers Edition
Practice Problems - Pg. 101

Adding Mixed Numbers with the Same Denominator

Solve these problems on a separate sheet of paper since it is often difficult to work out problems in a book.

1. $1\frac{3}{4} + 2\frac{1}{4}$

2. $1\frac{2}{16} + 1\frac{4}{16}$

3. $4\frac{1}{8} + 4\frac{3}{8}$

4. $3\frac{2}{7} + 6\frac{2}{7}$

5. $5\frac{1}{2} + 2\frac{1}{2}$

6. $6\frac{5}{11} + 12\frac{8}{11}$

7. $14\frac{14}{20} + 8\frac{3}{20}$

8. $15\frac{7}{13} + 1\frac{2}{13}$

9. $4\frac{3}{7} + \frac{2}{7}$

10. $10\frac{3}{5} + \frac{4}{5}$

Adding Mixed Numbers with Different Denominators

Solve these problems on a separate sheet of paper since it is often difficult to work out problems in a book.

1. $7\frac{1}{3} + 4\frac{1}{6}$ **2.** $2\frac{6}{11} + 5\frac{1}{33}$ **3.** $10\frac{1}{2} + \frac{6}{7}$ **4.** $11\frac{2}{3} + \frac{5}{8}$

5. $2\frac{5}{9} + 2\frac{7}{12}$ **6.** $6\frac{1}{20} + 10\frac{4}{5}$ **7.** $13\frac{2}{7} + 11\frac{1}{3}$ **8.** $1\frac{1}{3} + 4\frac{1}{2}$

9. $8\frac{2}{5} + 1\frac{4}{6}$ **10.** $7\frac{1}{2} + 6\frac{11}{13}$

Subtracting Fractions with the Same Denominator

Solve these problems on a separate sheet of paper since it is often difficult to work out problems in a book.

1. $\frac{2}{4} - \frac{1}{4}$
2. $\frac{8}{9} - \frac{6}{9}$
3. $\frac{2}{3} - \frac{1}{3}$
4. $\frac{3}{4} - \frac{1}{4}$
5. $\frac{7}{10} - \frac{4}{10}$
6. $\frac{4}{8} - \frac{1}{8}$
7. $\frac{4}{9} - \frac{1}{9}$
8. $\frac{10}{11} - \frac{6}{11}$
9. $\frac{11}{15} - \frac{2}{15}$
10. $\frac{9}{10} - \frac{7}{10}$
11. $\frac{4}{5} - \frac{2}{5}$
12. $\frac{4}{6} - \frac{1}{6}$
13. $\frac{6}{7} - \frac{5}{7}$
14. $\frac{7}{14} - \frac{3}{14}$
15. $\frac{5}{8} - \frac{1}{8}$
16. $\frac{8}{12} - \frac{4}{12}$
17. $\frac{15}{18} - \frac{3}{18}$
18. $\frac{12}{16} - \frac{4}{16}$
19. $\frac{12}{13} - \frac{6}{13}$
20. $\frac{1}{2} - \frac{1}{2}$

The Math Mechanic Series: Fractions & Mixed Numbers Edition
Practice Problems - Pg. 104

Subtracting Fractions with Different Denominators

Solve these problems on a separate sheet of paper since it is often difficult to work out problems in a book.

1. $\frac{1}{2} - \frac{3}{10}$

2. $\frac{5}{6} - \frac{7}{9}$

3. $\frac{14}{15} - \frac{5}{10}$

4. $\frac{3}{5} - \frac{5}{15}$

5. $\frac{8}{16} - \frac{5}{12}$

6. $\frac{5}{12} - \frac{5}{24}$

7. $\frac{4}{8} - \frac{4}{9}$

8. $\frac{2}{4} - \frac{1}{7}$

9. $\frac{3}{5} - \frac{3}{10}$

10. $\frac{17}{18} - \frac{1}{6}$

11. $\frac{9}{12} - \frac{5}{8}$

12. $\frac{13}{20} - \frac{2}{5}$

Subtracting Mixed Numbers with the Same Denominator

Solve these problems on a separate sheet of paper since it is often difficult to work out problems in a book.

1. $12\frac{1}{6} - 9\frac{1}{6}$ **2.** $4\frac{11}{25} - 1\frac{10}{25}$ **3.** $2\frac{20}{39} - 1\frac{5}{39}$ **4.** $1\frac{21}{30} - \frac{17}{30}$

5. $11\frac{29}{30} - 2\frac{27}{30}$ **6.** $4\frac{20}{27} - 3\frac{16}{27}$ **7.** $6\frac{10}{24} - 3\frac{5}{24}$ **8.** $1\frac{1}{2} - \frac{1}{2}$

9. $2\frac{7}{10} - 1\frac{3}{10}$ **10.** $11\frac{5}{6} - 1\frac{2}{6}$ **11.** $4\frac{3}{9} - \frac{6}{9}$ **12.** $10\frac{5}{8} - 5\frac{6}{8}$

Subtracting Mixed Numbers with Different Denominators

Solve these problems on a separate sheet of paper since it is often difficult to work out problems in a book.

1. $12\frac{4}{7} - 6\frac{1}{5}$ **2.** $7\frac{16}{18} - 4\frac{1}{9}$ **3.** $3\frac{8}{12} - 2\frac{14}{24}$ **4.** $9\frac{12}{16} - 8\frac{1}{6}$

5. $11\frac{2}{3} - 3\frac{6}{18}$ **6.** $7\frac{4}{6} - 2\frac{2}{4}$ **7.** $13\frac{1}{2} - \frac{3}{4}$ **8.** $8\frac{1}{4} - 7\frac{2}{3}$

The Math Mechanic Series: Fractions & Mixed Numbers Edition
Practice Problems - Pg. 107

Multiplying Fractions

Solve these problems on a separate sheet of paper since it is often difficult to work out problems in a book.

1. $\frac{1}{2} \times \frac{3}{8}$

2. $\frac{2}{5} \times \frac{3}{4}$

3. $\frac{4}{5} \times \frac{1}{3}$

4. $\frac{1}{2} \times \frac{1}{2}$

5. $\frac{1}{4} \times \frac{2}{3}$

6. $\frac{2}{6} \times \frac{3}{4}$

7. $\frac{5}{7} \times \frac{1}{6}$

8. $\frac{4}{8} \times \frac{3}{8}$

9. $\frac{1}{4} \times \frac{3}{4}$

10. $\frac{7}{12} \times \frac{10}{12}$

11. $\frac{3}{4} \times \frac{4}{5}$

12. $\frac{3}{7} \times \frac{6}{7}$

13. $\frac{2}{10} \times \frac{4}{7}$

14. $\frac{5}{11} \times \frac{2}{3}$

15. $\frac{5}{7} \times \frac{5}{7}$

16. $\frac{1}{10} \times \frac{1}{10}$

17. $\frac{27}{9} \times \frac{6}{18}$

18. $\frac{1}{2} \times \frac{4}{8}$

19. $\frac{14}{21} \times \frac{15}{7}$

20. $\frac{24}{8} \times \frac{24}{6}$

Multiplying Mixed Numbers

Solve these problems on a separate sheet of paper since it is often difficult to work out problems in a book.

1. $1\frac{1}{2} \times 3\frac{3}{4}$ **2.** $2\frac{2}{3} \times \frac{2}{6}$ **3.** $3\frac{1}{2} \times 1\frac{1}{2}$

4. $\frac{2}{4} \times 1\frac{5}{9}$ **5.** $\frac{10}{12} \times 3\frac{4}{6}$ **6.** $5\frac{4}{6} \times \frac{6}{8}$

The Math Mechanic Series: Fractions & Mixed Numbers Edition
Practice Problems - Pg. 109

Dividing Fractions

Solve these problems on a separate sheet of paper since it is often difficult to work out problems in a book.

1. $\frac{6}{7} \div \frac{24}{28}$

2. $\frac{2}{3} \div \frac{4}{6}$

3. $\frac{35}{25} \div \frac{15}{20}$

4. $\frac{3}{4} \div \frac{3}{4}$

5. $\frac{5}{7} \div \frac{8}{10}$

6. $\frac{8}{9} \div \frac{6}{12}$

7. $\frac{27}{60} \div \frac{18}{24}$

8. $\frac{7}{5} \div \frac{2}{3}$

9. $\frac{7}{8} \div \frac{21}{16}$

10. $\frac{12}{15} \div \frac{18}{20}$

The Math Mechanic Series: Fractions & Mixed Numbers Edition
Practice Problems - Pg. 110

Dividing Mixed Numbers

Solve these problems on a separate sheet of paper since it is often difficult to work out problems in a book.

1. $3\frac{1}{3} \div 2\frac{4}{5}$ **2.** $5\frac{1}{2} \div \frac{3}{4}$ **3.** $6 \div \frac{1}{4}$

4. $2\frac{2}{3} \div 1\frac{1}{3}$ **5.** $5\frac{5}{10} \div 10\frac{5}{10}$ **6.** $7\frac{1}{4} \div 4\frac{1}{2}$

Page 8 - Drill: Identifying Numerators and Denominators

1. 1 **2.** 6 **3.** 3 **4.** 8 **5.** 5

1. 8 **2.** 3 **3.** 7 **4.** 20 **5.** 10

Page 90 - Finding the Factors of a Number

1. 1

2. 1, 2

3. 1, 2, 4

4. 1, 2, 4, 8

5. 1, 3

6. 1, 7

7. 1, 2, 5, 10

8. 1, 5

9. 1, 2, 3, 6

10. 1, 3, 5, 15

11. 1, 2, 4, 8, 16

12. 1, 2, 4, 5, 10, 20

13. 1, 2, 5, 10, 25, 50

14. 1, 3, 9

15. 1, 2, 3, 6, 9, 18

16. 1, 5, 25

17. 1, 2, 3, 4, 6, 12

18. 1, 3, 11, 33

19. 1, 3, 9, 27

20. 1, 2, 3, 4, 6, 8, 12, 24

Page 91 - Finding the Greatest Common Factor

1. 3 **2.** 2 **3.** 5 **4.** 14 **5.** 7 **6.** 4 **7.** 6 **8.** 5 **9.** 11 **10.** 2

11. 3 **12.** 8 **13.** 5 **14.** 2 **15.** 9 **16.** 3 **17.** 4 **18.** 5

19. 2 **20.** 6

Page 92 - Converting Whole Numbers to Improper Fractions

1. $\frac{2}{1}$ **2.** $\frac{6}{1}$ **3.** $\frac{18}{1}$ **4.** $\frac{8}{1}$ **5.** $\frac{13}{1}$ **6.** $\frac{10}{1}$ **7.** $\frac{3}{1}$ **8.** $\frac{25}{1}$ **9.** $\frac{11}{1}$ **10.** $\frac{16}{1}$

11. $\frac{9}{1}$ **12.** $\frac{58}{1}$ **13.** $\frac{101}{1}$ **14.** $\frac{67}{1}$ **15.** $\frac{5}{1}$ **16.** $\frac{20}{1}$ **17.** $\frac{14}{1}$ **18.** $\frac{7}{1}$

19. $\frac{40}{1}$ **20.** $\frac{4}{1}$

Page 93 - Converting Improper Fractions to Whole Numbers

1. 5 **2.** 2 **3.** 2 **4.** 3 **5.** 2 **6.** 4 **7.** 3 **8.** 9 **9.** 9 **10.** 4

11. 2 **12.** 7 **13.** 2 **14.** 6 **15.** 5 **16.** 2 **17.** 5 **18.** 4 **19.** 11 **20.** 8

Page 94 - Converting Improper Fractions to Mixed Numbers

1. $1\frac{1}{7}$ **2.** $1\frac{1}{2}$ **3.** $1\frac{4}{5}$ **4.** $3\frac{1}{4}$ **5.** $2\frac{2}{8}$ **6.** $2\frac{5}{7}$ **7.** $1\frac{6}{14}$ **8.** $1\frac{4}{11}$

9. $1\frac{2}{5}$ **10.** $1\frac{14}{21}$ **11.** $1\frac{10}{15}$ **12.** $3\frac{6}{9}$ **13.** $1\frac{2}{4}$ **14.** $1\frac{7}{10}$ **15.** $1\frac{3}{6}$

16. $3\frac{1}{5}$ **17.** $1\frac{10}{12}$ **18.** $2\frac{6}{12}$ **19.** $5\frac{6}{7}$ **20.** $1\frac{1}{5}$

Page 95 - Converting Mixed Numbers to Improper Fractions

1. $\frac{9}{4}$ **2.** $\frac{7}{4}$ **3.** $\frac{20}{12}$ **4.** $\frac{17}{3}$ **5.** $\frac{27}{8}$ **6.** $\frac{11}{7}$ **7.** $\frac{21}{2}$ **8.** $\frac{24}{5}$ **9.** $\frac{12}{9}$ **10.** $\frac{19}{3}$

11. $\frac{17}{8}$ **12.** $\frac{46}{6}$ **13.** $\frac{34}{3}$ **14.** $\frac{9}{2}$ **15.** $\frac{52}{9}$ **16.** $\frac{15}{4}$ **17.** $\frac{83}{9}$ **18.** $\frac{31}{8}$ **19.** $\frac{13}{5}$ **20.** $\frac{94}{11}$

Page 96 - Reducing a Fraction to its Lowest Terms

1. $\frac{1}{2}$ **2.** $\frac{1}{3}$ **3.** $\frac{1}{3}$ **4.** $\frac{4}{5}$ **5.** $\frac{8}{11}$ **6.** $\frac{5}{9}$ **7.** $\frac{3}{8}$ **8.** $\frac{4}{11}$ **9.** $\frac{1}{5}$ **10.** $\frac{1}{2}$

11. $\frac{2}{3}$ **12.** $\frac{2}{9}$ **13.** $\frac{1}{6}$ **14.** $\frac{3}{7}$ **15.** $\frac{13}{21}$ **16.** $\frac{1}{4}$ **17.** $\frac{2}{5}$ **18.** $\frac{1}{3}$ **19.** $\frac{2}{5}$ **20.** $\frac{1}{4}$

Page 97 - Finding the Least Common Multiple

1. 12 **2.** 8 **3.** 12 **4.** 10 **5.** 18 **6.** 21 **7.** 8 **8.** 2 **9.** 10 **10.** 12

11. 7 **12.** 20 **13.** 18 **14.** 6 **15.** 30 **16.** 1 **17.** 24 **18.** 9 **19.** 14 **20.** 15

Page 98 - Finding the Lowest Common Denominator

1. $\frac{1}{4} + \frac{2}{4}$ **2.** $\frac{2}{6} + \frac{2}{6}$ **3.** $\frac{1}{2} + \frac{1}{2}$ **4.** $\frac{6}{8} + \frac{1}{8}$ **5.** $\frac{9}{15} - \frac{6}{15}$ **6.** $\frac{9}{24} - \frac{6}{24}$

7. $\frac{3}{10} - \frac{2}{10}$ **8.** $\frac{7}{10} - \frac{3}{10}$ **9.** $\frac{15}{20} + \frac{16}{20}$ **10.** $\frac{8}{18} + \frac{9}{18}$ **11.** $\frac{16}{24} - \frac{3}{24}$

12. $\frac{16}{28} - \frac{7}{28}$ **13.** $\frac{12}{18} - \frac{2}{18}$ **14.** $\frac{3}{9} + \frac{2}{9}$ **15.** $\frac{6}{27} + \frac{21}{27}$ **16.** $\frac{7}{28} - \frac{2}{28}$

17. $\frac{4}{12} + \frac{9}{12}$ **18.** $\frac{22}{38} - \frac{5}{38}$ **19.** $\frac{24}{30} + \frac{24}{30}$ **20.** $\frac{49}{56} - \frac{48}{56}$

Page 99 - Adding Fractions with the Same Denominator

1. $\frac{3}{4}$ **2.** $\frac{7}{8}$ **3.** $\frac{2}{5}$ **4.** $\frac{2}{3}$ **5.** $\frac{7}{9}$ **6.** $\frac{3}{4}$ **7.** 1 **8.** $1\frac{1}{5}$

9. $\frac{1}{2}$ **10.** $\frac{13}{20}$ **11.** $1\frac{1}{2}$ **12.** 1 **13.** $1\frac{1}{7}$ **14.** $\frac{2}{3}$ **15.** $1\frac{1}{4}$

16. $1\frac{1}{4}$ **17.** $\frac{1}{2}$ **18.** $1\frac{1}{2}$ **19.** 1 **20.** $\frac{3}{4}$

Page 100 - Adding Fractions with Different Denominators

1. $1\frac{1}{4}$ **2.** $1\frac{5}{12}$ **3.** $1\frac{3}{10}$ **4.** $\frac{3}{4}$ **5.** $\frac{17}{20}$ **6.** $1\frac{13}{18}$ **7.** $1\frac{11}{45}$ **8.** $\frac{3}{8}$

9. $\frac{23}{36}$ **10.** $\frac{3}{4}$ **11.** $1\frac{1}{22}$ **12.** 1 **13.** $1\frac{11}{18}$ **14.** 1 **15.** $\frac{3}{4}$ **16.** $1\frac{2}{15}$

17. $1\frac{11}{16}$ **18.** $\frac{8}{11}$ **19.** $1\frac{1}{5}$ **20.** $\frac{10}{21}$

Page 101 - Adding Mixed Numbers with the Same Denominator

1. 4 **2.** $2\frac{3}{8}$ **3.** $8\frac{1}{2}$ **4.** $9\frac{4}{7}$ **5.** 8 **6.** $19\frac{2}{11}$ **7.** $22\frac{17}{20}$ **8.** $16\frac{9}{13}$

9. $4\frac{5}{7}$ **10.** $11\frac{2}{5}$

Page 102 - Adding Mixed Numbers with Different Denominators

1. $11\frac{1}{2}$ **2.** $7\frac{19}{33}$ **3.** $11\frac{5}{14}$ **4.** $12\frac{7}{24}$ **5.** $5\frac{5}{36}$ **6.** $16\frac{17}{20}$

7. $24\frac{13}{21}$ **8.** $5\frac{5}{6}$ **9.** $10\frac{1}{15}$ **10.** $14\frac{9}{26}$

The Math Mechanic Series: Fractions & Mixed Numbers Edition
Answers to Problems - Pg. 114

Page 103 - Subtracting Fractions with the Same Denominator

1. $\frac{1}{4}$ **2.** $\frac{2}{9}$ **3.** $\frac{1}{3}$ **4.** $\frac{1}{2}$ **5.** $\frac{3}{10}$ **6.** $\frac{3}{8}$ **7.** $\frac{1}{3}$ **8.** $\frac{4}{11}$ **9.** $\frac{3}{5}$ **10.** $\frac{1}{5}$

11. $\frac{2}{5}$ **12.** $\frac{1}{2}$ **13.** $\frac{1}{7}$ **14.** $\frac{2}{7}$ **15.** $\frac{1}{2}$ **16.** $\frac{1}{3}$ **17.** $\frac{2}{3}$ **18.** $\frac{1}{2}$

19. $\frac{6}{13}$ **20.** 0

Page 104 - Subtracting Fractions with Different Denominators

1. $\frac{1}{5}$ **2.** $\frac{1}{18}$ **3.** $\frac{13}{30}$ **4.** $\frac{4}{15}$ **5.** $\frac{1}{12}$ **6.** $\frac{5}{24}$ **7.** $\frac{1}{18}$ **8.** $\frac{5}{14}$ **9.** $\frac{3}{10}$

10. $\frac{7}{9}$ **11.** $\frac{1}{8}$ **12.** $\frac{1}{4}$

Page 105 - Subtracting Mixed Numbers with the Same Denominator

1. 3 **2.** $3\frac{1}{25}$ **3.** $1\frac{5}{13}$ **4.** $1\frac{2}{15}$ **5.** $9\frac{1}{15}$ **6.** $1\frac{4}{27}$ **7.** $3\frac{5}{24}$ **8.** 1

9. $1\frac{2}{5}$ **10.** $10\frac{1}{2}$ **11.** $3\frac{2}{3}$ **12.** $4\frac{7}{8}$

Page 106 - Subtracting Mixed Numbers with Different Denominators

1. $6\frac{13}{35}$ **2.** $3\frac{7}{9}$ **3.** $1\frac{1}{12}$ **4.** $1\frac{7}{12}$ **5.** $8\frac{1}{3}$ **6.** $5\frac{1}{6}$ **7.** $12\frac{3}{4}$ **8.** $\frac{7}{12}$

Page 107 - Multiplying Fractions

1. $\frac{3}{16}$ **2.** $\frac{3}{10}$ **3.** $\frac{4}{15}$ **4.** $\frac{1}{4}$ **5.** $\frac{1}{6}$ **6.** $\frac{1}{4}$ **7.** $\frac{5}{42}$ **8.** $\frac{3}{16}$ **9.** $\frac{3}{16}$

10. $\frac{35}{72}$ **11.** $\frac{3}{5}$ **12.** $\frac{18}{49}$ **13.** $\frac{4}{35}$ **14.** $\frac{10}{33}$ **15.** $\frac{25}{49}$ **16.** $\frac{1}{100}$

17. 1 **18.** $\frac{1}{4}$ **19.** $1\frac{3}{7}$ **20.** 12

Page 108 - Multiplying Mixed Numbers

1. $5\frac{5}{8}$ **2.** $\frac{8}{9}$ **3.** $5\frac{1}{4}$ **4.** $\frac{7}{9}$ **5.** $3\frac{1}{18}$ **6.** $4\frac{1}{4}$

Page 109 - Dividing Fractions

1. 1 **2.** 1 **3.** $1\frac{13}{15}$ **4.** 1 **5.** $\frac{25}{28}$ **6.** $1\frac{7}{9}$ **7.** $\frac{3}{5}$ **8.** $2\frac{1}{10}$ **9.** $\frac{2}{3}$ **10.** $\frac{8}{9}$

Page 110 - Dividing Mixed Numbers

1. $1\frac{4}{21}$ **2.** $7\frac{1}{3}$ **3.** 24 **4.** 2 **5.** $\frac{11}{21}$ **6.** $1\frac{11}{18}$

Denominator: The bottom part of a fraction.

Factor: One of two or more expressions that are multiplied together.

Fraction: A fraction is part of a whole or group. A fraction is made up of a numerator and a denominator. The numerator is the number above the bar and the denominator is the number below the bar.

Greatest Common Factor: The largest number that will divide into two or more numbers evenly.

Improper Fraction: A fraction whose numerator is greater than or equal to its denominator.

Least Common Multiple: The smallest number that is the multiple of two or more numbers.

Lowest Common Denominator: The smallest multiple of the denominators of two or more fractions.

Minuend: A value to be subtracted from.

Mixed Number: A combination of a whole number and a fraction.

Multiple: Product of a particular number and any other number.

Numerator: The top part of a fraction.

Quotient: The result obtained when dividing one quantity by another.

Remainder: The number left over when one number is divided by another.

Repeating Decimal: A decimal in which the digits endlessly repeat in a pattern.

Subtrahend: A value that will be used to subtract.

About The Math Mechanic Series

The Math Mechanic does the dirty work that textbooks and other supplementary books don't. It shows the step-by-step mechanics of solving mathematical problems. To survive in today's world it is imperative that you have the ability to solve mathematical problems. Whether it is working out problems for schoolwork or performing calculations related to money, strong mathematical skills are a necessity. Our books are easy to understand guides that are designed to eliminate the frustration that many Children and Adults have with mathematics. Each book is a combined Tutorial, Workbook, and Reference Manual. From Students to Parents to Professionals, The Math Mechanic Series provides the most potent and comprehensive set of tools for helping anyone to become proficient with mathematics.

About The Authors

The books of The Math Mechanic Series were written by private tutors of The Math Mechanic private tutoring service. Our tutors took notice of the hundreds of requests for a book that was a combined tutorial, reference manual, and workbook which also contained the procedures for solving mathematical problems with a plethora of fully worked step-by-step examples. Our tutors implemented the above and added many other learning tools to make this book one of the most powerful and easy to comprehend math books that can be found in bookstores.

Printed in the United States
998700002B

9 781593 300340